ENGLISH HUNGER
AND INDUSTRIAL
DISORDERS

William Wildman, Viscount Barrington
(1717–93)

ENGLISH HUNGER AND INDUSTRIAL DISORDERS

A study of social conflict during the first decade of George III's reign

WALTER J. SHELTON

Associate Professor of History
Wilfrid Laurier University

University of Toronto Press

First published 1973 in the United Kingdom by The Macmillan Press Ltd
First published 1973 in Canada and the United States by
University of Toronto Press, Toronto and Buffalo

ISBN 0–8020–2087–9

Printed in Great Britain

TO ROSALEEN

Contents

List of Maps

List of Illustrations

Preface

It is impossible to write on English rioting of the eighteenth century without incurring many debts. The extent to which I have drawn on the works of Dorothy George, George Rudé and E. P. Thompson will be quite apparent.

My own interest in the subject was stimulated by Professor John M. Norris of the University of British Columbia, without whose encouragement and guidance this work would never have been completed.

The assistance of the staffs in most of the County Record Offices in southern England, especially in East Suffolk, Gloucestershire, Middlesex and Norwich, was invaluable. I also owe a debt of gratitude to the House of Lords Record Office, the British Museum, the Public Record Office, the William Clement Library in Ann Arbor, Michigan, and the Sheffield Central Library.

At Wilfrid Laurier University I have been fortunate to have the advice and support of my colleagues; in particular Mrs Agnes Hall who worked on two of the maps and Jim Harkins who, with his knowledge of French hunger riots of the 1790s, has helped to focus some of my ideas.

The debt I owe to my wife and children cannot adequately be expressed.

W.J.S.

Introduction

This work deals with riots in the first decade of George III's reign. By focusing both on the immediate causes of these disturbances and on the underlying social tensions which determined their form and direction, it seeks to explain why this was one of the worst periods of disorder in the century. Early studies of the riots of the 1760s which have not dealt exclusively with political disturbances have treated many of the riots as part of the history of trade unions or of the story of the rural labourer's degradation. As a result, the interrelationship of the provincial hunger riots and the metropolitan industrial riots has been ignored by most historians of popular movements. More recent studies have presented these expressions of social tension primarily in terms of the discontents of the rioters. By focusing closely upon the 'faces in the crowd' scholars have corrected the misconception that eighteenth-century mobs were chiefly composed of the most depraved elements in society. But in the process of this legitimate attempt to rehabilitate the historical 'crowd', such students have been rather reluctant to concede the manipulation of rioters by those outside the mob. This is particularly true when rioters clearly acted according to socially appropriate goals, as was usually the case with rural hunger mobs and industrial strikers. The result of such emphasis on the rioters is the undervaluing of the role of other interests, and the stressing of immediate at the expense of secondary causation.

I have attempted in the following pages to set the rioters of the 1760s in their social context and present the riots as the product of an interaction of the poor, the landowners, the industrialists, the local authorities, and the national government. All of these interests contributed to disorder in some fashion: by suggesting the poor regulate markets for themselves, the gentry encouraged them to take actions for which many later were tried by special assize; by failing to suppress the initial disorders, the

1

magistrates appeared to sanction the acts of the mobs; by blaming middlemen for high prices of food, clothiers and other industrialists in the distressed cloth counties of southern England diverted their underpaid workers towards bunting mills and local markets; by proclaiming the old anti-middlemen statutes against forestalling, engrossing, and regrating instead of ending grain exports, the Ministry confirmed that the food shortage was artificial and encouraged further attacks upon middlemen and farmers; by blaming coal-undertakers and then failing to enforce existing legislation against these middlemen of the coal trade, the government encouraged coalheavers to act in their own defence.

While the distress of the industrious poor, which followed sudden fluctuations in food prices and declining employment, was the common denominator of the numerous riots of the 1760s, such disorders were merely the surface manifestations of underlying political, economic, social, and intellectual ferment, which affected all levels of society. This state of extreme flux influenced the actions and attitudes of the upper and middling sorts, as well as those of the lower orders. The riots were the product of an interaction of various interests, especially of the industrious poor on the one hand and the ruling orders on the other, rather than merely the expression of the discontents of one group. The responses of the poor and the privileged alike can only be explained with reference to important social changes, which resulted after the mid-century from agricultural and industrial developments. The effect of these social changes was aggravated by war and by the progressive abandonment of the principles and practices of the old 'moral economy'.

After the mid-century, the grievances of the poor were more deeply seated than could be accounted for by resentment at spiralling prices, low wages, scarce employment, or the tyranny of a Ministry towards John Wilkes, the champion of the 'rights of free-born Englishmen'. Their actions must be seen against a background of social tension among the industrious poor. This background however provides only a partial explanation of the riots of the first decade of George III's reign. In all disorders there is an interaction between the authorities and the rioters; neither can be considered in isolation. The actions of the local and national authorities were frequently equivocal and require explanation.

Not only the poor but also the more privileged orders of society were caught up in the changes which confused and frightened them by the 1750s. After the mid-century, the growth of commercial farming, which was stimulated by the expansion of urban populations and military victualling contracts, disturbed the social balance of the countryside.[1] Large farmers and middlemen emerged as the chief beneficiaries of agricultural growth and their social pretensions appeared to threaten the leadership of many of the lesser parish gentry. The militancy of returned veterans of the Seven Years' War, whose expectations about civilian life had changed with service abroad, threatened the stability of rural and urban society. Men trained in the militia, the army, the navy, and Irish terrorist gangs played a critical role in the riots by providing a disciplined core of militants able to defy the military, and by giving direction to the disorders. In times of extensive disturbances, the landed interest, together with leaders of industry and commerce, feared the intentions of a large proportion of the population. General insurrections were always a nightmare possibility in a century when the sparse forces of order looked pitifully weak in the face of any serious threat to the social order. Although suspicion of standing armies remained throughout the century, distrust of the militia caused a growing reliance upon the regular army in the face of social protests. In times of unrest, the ruling orders often anticipated the outbreak of serious disorders and resorted to the tactic of diverting the disaffected against selected scapegoats. In doing this they encouraged rioters to violence and helped to shape events. This ploy was particularly evident in the hunger riots of 1766, but both the authorities and employers also used it in the industrial disturbances of 1768.

Serious rioting in the 1760s was the means whereby English society sought to achieve radical change as it moved towards a new equilibrium. This decade saw the beginning of a period of social transition which stretched into the next century. In these ten years, many of the changes which all classes were to feel acutely sent shock waves through English society. Because they followed several years of relative improvement for the poor and because they developed so rapidly, the pressures of the 1760s provoked stronger than usual protests from the dispossessed. As

[1] See J. D. Chambers and G. E. Mingay, *The Agricultural Revolution 1750–1880* (London: B. T. Batsford Ltd, 1966).

one observer noted, the poor were 'too much oppressed, and the burthen of late years [had] come too fast upon them to be born with patient resignation'. Later in the century, the social problems emerging at this time became more serious without prolonged disorders, because the pressure on the poor built up steadily over a period so that they had time to adjust their expectations.[2] Although the prices of food were higher and economic dislocation more severe at other times, riots were more serious and sustained in the 1760s because of the frustrations of both the ruling orders and the poor.[3]

Over forty years after Sir Lewis Namier first published his analysis of the structure of their politics, the early years of George III's reign continue to attract the attention of historians. Students of political reform have seen the 1760s as a period of transition from political agitation and rioting to more formally organised reform movements.[4] That this decade was also a period of vigorous social protest has only become apparent in more recent times.

In the 1760s, serious riots occurred in provincial and metropolitan districts of England. Because such disorders were frequently the work of one level of society, the industrious poor, they greatly alarmed the upper and middling sorts of the kingdom. One correspondent expressed the view of most of the ruling orders when he observed : 'These are very unhappy times; but when they will mend or how they can be mended, I am at a loss to conceive; for they have let the minds of the lower people take too strong a bias to anarchy, for want of being stopped in time.'[5] Usually the distress produced by a combination of high prices and reduced income was the precipitating cause of the outbreaks, or in the case of the Wilkite political disorders, was not far in the background. In these years tailors, coopers, watermen, shoemakers, colliers, tinners, seamen, coalheavers, farm-workers, domestic servants, and others of the lower orders vigorously demonstrated their social discontents through the only means

[2] *Gazetteer and New Daily Advertiser*, 1 November 1766.
[3] Prices of wheat and bread were highest for the century in 1795 and 1799. Thomas Southcliffe Ashton, *Economic Fluctuations in England 1700–1800* (Oxford : Clarendon Press, 1959) p. 181.
[4] See I. R. Christie's Introduction to George Stead Veitch, *Genesis of Parliamentary Reform* (London : Constable, 1964).
[5] Mw. Fetherstonhaugh to Newcastle, 7 June 1768, British Museum, Additional MSS., 32990, fol. 180–1.

available to them, the riot. In doing so, they appeared to many to threaten the social order. Nor were men the only ones to demonstrate violently. One observer wrote of three hundred women, lawn-clippers, marching in white in Maxweltown, Scotland, escorted by crowds of journeymen weavers and others.[6] Even the 'ladies of pleasure' of the Metropolis rioted over the exorbitant demands of bawds, pimps, tavern-keepers, and waiters. One London newspaper reported their dispute in the following terms :

They ground their hardships (and indeed with some show of justice) upon the same foundation, and almost in the same terms, with the *coalheavers*, viz. that the *inordinate burdens* they *lie under*, and which they so often *bear* and *groan* with for three parts of the year together, more or less, wear them out so soon, that unless their wages are doubled or they have some *settlement* (which they would like better) they must be obliged, when they are *battered with labour*, either to turn overseers to the younger part of the profession, vulgarly called bawds, or retire from the world as *penitent prostitutes*.[7]

Unfortunately the records do not always reveal the outcome of such intriguing labour disputes, but their number and variety indicate the tensions of a society in transition.

In one sense all major disturbances in eighteenth-century England are political because to a greater or lesser extent they threatened the dominance of the ruling order. For the sake of convenience, however, the major riots of this decade may be grouped into three categories according to their primary goals: agrarian hunger riots which developed in the market towns and rural parishes of southern England; industrial riots related to the labour disputes of seamen, coalheavers, silk-weavers, and other groups of England's labouring poor; and political riots centring around the causes and person of John Wilkes. Riots, however, frequently overlapped, and defied such tidy categorisation. Rioters, once set going, often addressed themselves to the correction of more than the one grievance which had precipitated their protest. Farmers and labourers, who began by seizing militia muster sheets from magistrates, later complained of the aristoc-

---

[6] *St. James's Chronicle*, 24 May 1768.
[7] Ibid., 17–19 May 1768.

racy and gentry enjoying their land for too long;[8] coalheavers rioting over the extortionate demands of coal-undertakers shouted for 'Wilkes and Liberty' and joined political demonstrations against the Ministry;[9] yeomen and labourers, having pulled down partly-constructed houses of industry in East Anglia, vowed to lower the prices of provisions in neighbouring markets;[10] seamen, demonstrating before Parliament for higher wages, cheered for the King and drove off Wilkite supporters;[11] or hunger rioters in Jersey, having successfully forced down food prices, vigorously pressed political reforms upon their reluctant governing council.[12]

For its part, the government often read into the blurring of distinctions between riots more than they warranted. The mere incidence of rioting at home and abroad in this period was sufficient for the Ministers to connect them together and to weave conspiracy theories in an age when the prejudice of the governing classes against standing armies was strong and the existing forces of order weak. In the climate of disorder of this decade, both at home and overseas, the attitude of influential interests in English society to social and political change hardened.

The precise relationship between riots in England, Ireland, and America awaits the definition of historians. Certain connections are evident. American radicals hailed John Wilkes as an ally and utilised his causes for their own purposes. They were encouraged in this by English protest movements. In a letter to the House of Assembly of South Carolina, for example, the Bill of Rights Society stressed their mutuality of interests in the defence of common rights for 'Property is the Natural Right of mankind, the connection between taxation and representation is its necessary consequence. Our case is one, our enemies the same.'[13] Many seamen who played a leading role in the pre-revolutionary dis-

---

[8] John R. Western, *The English Militia in the Eighteenth Century*, Studies in Political History, ed. Michael Hurst (London: Routledge & Kegan Paul, 1965) p. 300.

[9] George Rudé, *Wilkes and Liberty* (Oxford: Oxford University Press, 1962) p. 97.

[10] Public Record Office, *State Papers*, SP 37/4, fol. 196/595 and fol. 202/595.

[11] Seamen were not consistent in their political loyalties. On another occasion a large body of sailors reportedly escorted Wilkes across London Bridge to Westminster. See *St. James's Chronicle*, 7 May 1768.

[12] *Calendar of Home Office Papers* (1766–69) 528–33, no. 1361.

[13] Public Record Office, *Chatham Papers*, PRO 30/8/56, fol. 96–7.

orders in Boston probably were familiar with the political and social protests in England, and may even have participated in them.[14] Weavers in London and Dublin exchanged information on wages and tactics to be used against their masters.[15] As a result of heavy waves of immigration and transportations, many of the more alienated of British lower-class society must have swelled the ranks of the American disaffected in the decade before the War of Independence.[16] Another source of conflict between the revolutionaries and the authorities in America may well have been the fact that regiments such as Burgoyne's Light Horse which had suppressed riots in England of the 1760s later served in the American campaigns between 1775 and 1783. More important than the real connection of events at home and abroad was the apprehension of such a connection by influential interests. The privileged orders of English society saw a clear relationship between riotous events in England, Ireland, and America. As one politician sardonically observed, 'Has not the mob of London as good a right to be insolent as the unchecked mob of Boston?'[17] The rulers of England perceived a common decline towards anarchy which was to be opposed in all the King's dominions.[18] Hence they adopted increasingly rigid attitudes towards mobs at home and abroad.

Although class awareness among the labouring poor in its fullest sense had to await the full development of the factory system in the next century, there was emerging a new polarisation of class attitudes in the 1760s.[19] While often in this decade disputes were between interest groups within the ranks of the industrious poor, conflicts between industrial workers and owners over

[14] L. Jesse Lemisch, 'Jack Tar Versus John Bull, The Role of New York's Seamen in Precipitating the Revolution' (unpublished Ph.D. dissertation, Yale University, 1962).

[15] *Calendar of Home Office Papers* (1766–69), no. 1317, 20 October 1769.

[16] Ashton, *Economic Fluctuations in England 1700–1800*, p. 159; see also Mildred Campbell, 'English Emigration on the Eve of the American Revolution', *American Historical Review*, LXI, no. 1 (October 1955).

[17] Mr Wedderburn to Mr Grenville, 3 April 1768, *Grenville Papers*, ed. William James Smith, 4 vols (London: John Murray, 1853) IV 263–5.

[18] Barrington to Adam Jellicoe, 1 September 1768, Ipswich and East Suffolk Record Office, Ipswich, *Letter Book of Viscount Barrington, Barrington Papers*.

[19] Asa Briggs, 'The Language of Class', in *Essays in Labour History*, ed. Asa Briggs and John Saville, Papermac (London: Macmillan, 1967).

wages and conditions occurred more frequently than before and revealed the dawning of class identity. With the progressive abandonment of the principles and practices of the old 'moral economy', workers found it useless to direct their energies solely towards lowering the prices of 'necessaries'. They at first demanded the application of the old protective statutes regulating food prices, wages, apprenticeship and foreign competition. When for the most part the authorities failed to do what they wished, the industrial poor rioted and struck for higher wages and better conditions. While traditional goals and tactics of the poor continued to exist alongside the more novel ones, the first decade of George III's reign was a transitional one in labour relations.

This transition was more apparent in the strikes and riots among both the provincial and the metropolitan industrial workers in 1760s than among agricultural labourers and reflected the emergence of embryonic trade unions. As the Webbs noted, eighteenth-century trade unions were distinct from earlier craft guilds.[20] They appeared at the beginning of the century among workers in the cloth trade, which was among the first trades to be operated on a capitalistic basis. The organisation and goals of the woolcombers, weavers, and others of the woollen cloth industry, which were quickly imitated by other industrial groups, were the result of the separation of the workers from the ownership of the capital and machinery essential to the production of cloth goods. By the 1760s many of the journeymen in London trades had come to recognise the improbability of their rising to positions of ownership, and adjusted their organisations and goals to suit their economic and social expectations. The relatively sophisticated tactics and demands of the provincial and metropolitan industrial workers, which were evident in the 1760s, reflected the slowly-emerging industrial society.

Although it extended through several social orders and was by no means as cohesive as some commentators suggested, the landed interest was more homogeneous in outlook, especially in economic affairs, than the industrious poor, who were not only divided into finely-graded social orders, but also multitudinous economic

---

[20] Sydney and Beatrice Webb, *The History of Trade Unionism* (1894), Reprints of Economic Classics (New York: Augustus M. Kelley, 1965) p. 16.

interests.[21] Similarly, the rising industrialists had a sense of identity as a distinct interest group. The effect of the disorders of the 1760s upon these two relatively homogeneous interests was to increase their sensitivity to the threatened 'tyranny of numbers'. Thereafter, despite their mutual dislike and suspicion, both these groups in times of disorder stood together against the threat from below, of which they first became really aware in the 1750s and 1760s.

The focus of this work will be the hunger and industrial riots of 1763–9. These riots and the Wilkite disorders, while related through a common background of social unrest and economic deprivation, are better treated separately. For reasons which will be discussed in Part Two below, the Wilkite disorders will be peripheral to this study.

In analysing the background to the events of the 1760s, it will be necessary to consider developments earlier in the century, especially those of the previous decade. Some understanding of the evolution of attitudes of various groups to farmers and middlemen, for example, is an essential prerequisite to explain the extraordinary hostility of both the authorities and the poor to these two interests in the hunger riots of 1766–7. Similarly, the polarisation of class interests over the Militia Act crisis of 1757 and the food riots connected with it help to explain the equivocal role of the authorities in 1766–7 and the apparent malleability of the poor.

The paucity of detailed source materials hampers the student of the English mobs of the eighteenth century.[22] The absence of a national police force and the lack of an efficient bureaucracy largely account for such deficiencies. Calendars of prisoners delivered to the four special assizes held in Wiltshire, Berkshire, Gloucestershire, and Norfolk in December 1766 rarely indicate more than the names and alleged offences. Other Treasury Solicitor's papers such as indictments held at the Public Record Office

---

[21] M. Dorothy George, *London Life in the Eighteenth Century* (London: Kegan Paul, Trench, Trubner & Co. Ltd, 1925; reprinted, New York: Harper Torchbooks, 1965) *passim*.

[22] George Rudé has contrasted the detailed records kept by the French police in this century. George Rudé, *The Crowd in History, A Study of Popular Disturbances in France and England, 1730–1848* (New York: John Wiley & Sons, 1964) p. 13.

are equally vague. Sessional record books, collections of depositions, and prosecution briefs are more valuable sources of detailed
information about the composition of mobs, but are rarely available. Only Norwich Record Office appears to have these types of
material for 1766–7, and one may write with more confidence
about the social complexion and goals of the rioters who did
great property damage in that city in late September 1766,
than about other hunger rioters of the period. Generally, details
of age, occupation, military attachment, physical characteristics
of rioters and other information essential for a close analysis of
riotous mobs are only recorded in descriptions of suspects believed
to have absconded.[23] Marching Orders of the Army, newspaper
accounts, and scattered private correspondence only partially fill
the gaps in extant legal documents.

It is now almost a truism to remark at the beginning of any
study of rioting that most records are subject to the distortions of
the class sympathies of the officials and the articulate minority
who wrote of the events in documents, newspapers, and letters
to friends.[24] By definition, the inarticulate poor left few records
of their motivations and actions. Their story has to be pieced
together from indirect sources, always remembering that the
authorities pursued a deliberate policy of discrediting them. In
the 1750s, for example, Lord Mansfield advised the Marquis of
Rockingham, the Lord Lieutenant of the West Riding, to publi-
cise widely the dishonesty of the rioters in order to discredit them
in the eyes of the public by having sheriffs send advertisements
across the country to show the 'wicked spirit that has blown up
the mob, the instances where they have ended in exacting money
in theft or robbery. Undervaluing them that they ought to be
suppressed by the Civil Magistrates to leave the military force
free to attend to matters more important.'[25] There can be little
doubt that the same policy was pursued a decade later when
more serious riots broke out. Early accounts of food riots in 1766
commented on the honesty of mobs, who forced the sale of grain

[23] See Norwich Record Office, *Norwich Quarter Sessions Order Book*
(1755–75).
[24] E. P. Thompson, *The Making of the English Working Class* (London:
Victor Gollancz, 1963) p. 59 and *passim*.
[25] Mansfield to Rockingham, 4 October 1757, Central Library, Sheffield,
Rockingham MSS., R1–108.

at 'just' prices and saw that the former owners received their money from the sales. One astute observer noted the 'Mob was honest but resents any fraud on itself'.[26] Later accounts report growing property damage and looting. This changing picture may therefore reflect the growing exasperation of the rioters, increased alarm among the privileged classes at the earlier disciplined restraint of the disaffected whom they now sought to discredit, or a combination of both.

Newspaper publishers, sensitive to the criticism that detailed reporting of riots stimulated disturbances elsewhere, provided incomplete news coverage of disturbances. In October 1766, for example, the *Public Advertiser* published a letter of complaint from one subscriber who accused the editor of inciting unrest by constantly reporting riots in detail. The editor acknowledged the danger, and thereafter few accounts of riots appear in that paper for the remaining months of 1766. The magistrates of Norwich actually ordered the local newspapers to refrain from including graphic details after two days of extensive riots in September 1766 to avoid inciting further outbreaks.

By order of the magistrates. Printing and publishing in newspapers, the various excesses of rioters and disturbers of the peace, being little less than holding out examples for the wicked and profligate in other places, it is thought sufficient to acquaint our readers, that from specious pretences a great number of the lowest people wantonly destroyed provisions in the last Saturday's market, and committed other outrages.[27]

There was perhaps some justification for such censorship because newspapers were prone to publish rumours which circulated through the market places and gave exaggerated accounts of disturbances. One writer reported three examples of distortions from western England: (1) The mob removed flour from Mr Cambridge's mill at Whitminster with little damage. Lord Hardwicke's report spoke of a mill pulled down in Cambridgeshire; (2) A nervous man ran at the sound of the Cow's Horn, the signal of the mob, and his wife had fits. Four miles away in Gloucester the report spread that his house had been 'pulled

---

[26] John Pitt to Hardwicke, 29 September 1766, Add. MSS., 35607, fol. 290.

[27] *Norwich Mercury*, 4 October 1766, datelined 2 October.

down about his ears and his ricks destroyed'; (3) Dragoons who
had been sent to Cirencester were reported all killed. An officer
who inquired into the situation reported to the War Office that
many had been wounded and seven killed. Four days later all
the soldiers returned to Gloucester, a mere ten miles away.[28]

Even if the records of trials were complete and accurate, one
could not be certain that those arrested for rioting were a cross-
section of the mobs. Magistrates were often unable to arrest
rioters for several days or even weeks after their alleged offences.
Without military assistance the authorities were unable to make
summary arrests. As the army gradually restored calm to the
rural areas in October and November 1766, the magistrates concentrated their efforts on hunting down the ringleaders. Finding
witnesses, taking depositions, and finally tracing offenders were all
time-consuming, and the justices frequently enlisted the aid of
churchwardens and other parish officials. They asked for information about the names of known rioters, villagers who were
absent from their home parishes during the riots, or people who
had subsequently absconded.[29] Consequently, those indicted for
rioting were almost invariably locals.[30] Despite their absence from
lists of prisoners, one may suspect that certain outside groups did
in fact participate in the disorders. Similarly, although documents rarely identify prisoners as ex-servicemen, the tactics of the
mob, the dress of the participants and the comments of the
authorities strongly hint at the importance of the veterans of
the Seven Years' War in the disturbances of the 1760s.[31]

The dangers of interpreting eighteenth-century statistics are too
well known to historians to require lengthy comment here.[32]
Those relating to the prices of provisions, particularly grain, are
most relevant to this work, because price fluctuations correlate
closely with the incidence of hunger riots. Broadly, there are two
sources for such figures: certain wholesale price records kept by
such institutions as Eton College, and monthly lists of market
prices sporadically published by the *Gentleman's Magazine* and

[28] Pitt to Hardwicke, 29 September 1766, Add. MSS., 35607, fol. 290.
[29] *Norwich Quarter Sessions Order Book* (1766).
[30] Cf. George Rudé, 'The London Mob of the Eighteenth Century',
Historical Journal, II, no. 1 (1959) 1–18.
[31] See Part One, Chapter 4 below.
[32] See Sir George Norman Clark, *Guide to English Commercial Statistics
—1696–1782* (London: Offices of the Royal Historical Society, 1938).

other journals. For the establishment of short-term fluctuations in the cost of living, the wholesale prices are less valuable than the retail market prices. Usually based upon average prices at two seasonal dates in the year, Lady Day and Michaelmas, they bore only an indirect relationship to the prices of the food that the labouring poor were forced to buy in the market place or in the bakers' shops.[33]

Although the prices listed in the press are more valuable in studying the precipitating causes of popular protests in the eighteenth century, this source has its disadvantages too. The publishers of the *Gentleman's Magazine*, for example, gleaned information on local food prices only with considerable difficulty through the aid of voluntary correspondents.[34] Their varying degree of detail reflected current public concern over comparatively short crisis periods. In 1767 the *Gentleman's Magazine* attempted to record accurate market statistics for various divisions of England in the interests of obtaining more rational grain legislation, but the unreliability of its voluntary correspondents soon forced the abandonment of this attempt.

Even when such information was forthcoming, it was disappointingly vague and incomplete. Usually newspapers and journals published a range of prices at which grain sold in local markets on a given day. They rarely distinguished between grades of grain. There were, for example, in the summer of 1766 two ranges of prices, one for old and one for new grain, and within each range there were two levels, the price paid by dealers which was negotiated in private, and the 'pedling price' which was paid in the open market and was higher.[35]

Nor was the fact that grain sold at a particular price in a local market according to the monthly figures in the *Gentleman's Magazine* any guarantee that it could be purchased by anyone at that price. Barley and oats in southern England, for example, usually went directly to the brewers, distillers or other dealers

---

[33] Many eighteenth-century polemicists used the wholesale prices of Eton College. See *Considerations on the Exportation of Corn* (anonymous pamphlet, 1766).

[34] *Gentleman's Magazine*, xxxvi (1766); xxxvii (1767); xxxviii (1768).

[35] Charles Townsend to Grafton, 4 September 1766, West Suffolk County Record Office, Bury St Edmunds, *Grafton Papers*.

rather than to the retail market.[36] This partly explains why the prices of the coarser grains throughout the riot months remained curiously steady and well below the price of wheat.

Although there are strong indications that prices of food of the populace did spiral upwards in times of social discontent and one can establish a correlation in most instances, there are anomalies. The fact that prices rose sharply in parts of the country like Wales without disorders occurring, while elsewhere prices remained stable in centres like Worcester where riots still took place, raises interesting questions.[37]

Eighteenth-century grain statistics are at best imprecise measurements of discontent. They are broad indicators of the conditions of the poor. Even when they appear to be reasonably accurate and complete, they must be related to the prices of other provisions, to wages and employment possibilities, and the expectations of the people. But a knowledge of the attitudes of the poor and their reactions to their conditions is itself insufficient to determine the causes of riots. The responses of the poor do not exist in a vacuum. One can focus too narrowly on the 'faces in the crowd'. The attitudes of other significant groups and their roles in the events of 1766–8 are vitally important.

Because of the present limitation of source materials, any study of the riots of the 1760s must remain chiefly impressionistic. Patient culling of county record offices by teams of researchers may yet uncover adequate material for the essential local studies of eighteenth-century riots. To date, historians have only begun this type of work on riots of the next century. When a whole series of studies comparable to the one produced by A. J. Peacock on the East Anglian agrarian riots of 1816 appears, a new work of synthesis may proceed.[38] Work on eighteenth-century riots appears likely then to proceed in two directions: local studies as noted above, and comparative studies of American, Irish, and

---

[36] See *Gentleman's Magazine,* xxxvi (1766) and xxxvii (1767); and House of Lords Record Office, Westminster, *Committee on High Prices of Provisions* (March, 1765) *Main Papers.*

[37] *Gentleman's Magazine,* xxxvi (1766). See monthly grain prices for summer and autumn months.

[38] A. J. Peacock, *Bread or Blood: A Study of the Agrarian Riots in East Anglia in 1816* (London: Victor Gollancz, 1965).

English disorders.[39] The links already apparent are suggestive of interesting insights into the social developments of the three countries.

Meanwhile, although conclusions must perforce be tentative, it seems appropriate to undertake a re-examination of the disorders of the 1760s. Such an impressionistic study has value because pioneer, semi-sociological analyses of the historical mob in Britain and France raise questions about the wider context in which the English rioters of the eighteenth century operated.[40] Is it valid to explain the actions and goals of rioters solely or even predominantly in relation to the 'faces in the crowd'? If the fluctuating prices of provisions and the apparent threat of starvation precipitated them, did the responses of the authorities and other influential groups affect the form and direction of the riots? Were in fact even rural hunger mobs as free from the influence of those standing outside and apart from them as George Rudé would perhaps have us believe?[41] He is more willing to see urban political mobs, rather than hunger rioters, manipulated by those standing apart from them. If the actions and goals of the mobs were consistent with their social composition, why did they not attack equally appropriate targets that went unmolested? More specifically, if they attacked corn dealers and large farmers, why did they not assault industrialists, such as coal owners and clothiers who kept down wages, or gentry with whom they had recently quarrelled over discriminatory militia legislation? Are influential interests whom the rioters did not attack as important in estimating motivation as those whom they did attack? This abstention from rioting against employers in 1766 is remarkable, given the economic recession of the late 1760s and the history of industrial tensions. Clothing workers had engaged in bitter disputes over wages with West Country clothiers less than a decade earlier;[42] even more recently coal miners had attacked the property of their masters to gain concessions in working con-

[39] See Pauline Maier, 'John Wilkes and American Disillusionment with Britain', *William and Mary Quarterly*, xx, 3rd ser. (1963) 373–95.

[40] Georges Levebvre, George Rudé, *et al.*

[41] Rudé, *The Crowd in History*.

[42] Elizabeth Waterman Gilboy, *Wages in Eighteenth-Century England*, Vol. xlv of *Harvard Economic Studies* (Cambridge, Mass., Harvard University Press, 1934) p. 80 *et seq.*

ditions and to maintain wage levels.[43] Why did not the industrial workers demand higher wages and continuous employment from their employers as the cost of living rose? Can the actions of the various industrial mobs, such as seamen and coalheavers, be explained without writing of their exploitation and manipulation by government-paid leaders as well as private enterprisers? Consideration of these and other important questions will be given in subsequent chapters.

All students of the social structure of eighteenth-century England face problems of definition. While more precise explanations will be provided on an *ad hoc* basis in footnotes and text, some general comment is required by way of introduction. Asa Briggs has noted that people living in this period did not use the term 'class' to describe social groupings, and preferred to employ names such as 'interests' or 'orders'.[44] In this study the phrase 'the landed interest' will frequently appear and will distinguish all who gained income from renting or working their own land. In this economic sense, the name applies to the great landowners who rented or developed land, as well as to lesser gentry and yeomen who farmed part or all of their land. Socially the landed interest extended through several orders; from the aristocratic landowner down to the yeoman. The term embraces widely different economic and social levels. Only the rentier connection with the land gives unity.

Certain other eighteenth-century terms are confusing. The name 'yeoman' originally had a fairly precise meaning, and was applied to the order below the gentry and above the common farmer or peasant. During the course of the eighteenth century, this term came to have a much wider application. There were yeomen carpenters and yeomen weavers, to name but two, as well as yeomen farmers. Perhaps a modern parallel for such a development is the title 'esquire' which also has lost its precise meaning long ago. In this study 'yeomen' will be used in its older sense of 'better-class' farmer.

Perhaps the most common phrases for the upper levels of English society were the 'upper and middling sorts'. These terms are as imprecise as modern class categories, but like them they

---

[43] *Gentleman's Magazine*, xxxv (1765) 430, 488; *State Papers*, SP 37/4, fol. 1.
[44] Briggs, 'The Language of Class'.

are useful when precise descriptions are impossible, for they do convey some meaning. Generally in urban centres the middling sort refers to that large group of well-to-do merchants, artisans, craftsmen, and the rest who lay between the aristocratic and sub-noble financial and commercial interests on the one hand and the vast conglomeration of the industrious poor on the other. This latter term refers to the labourers, weavers, manufacturers, seamen, miners, porters, watermen, and many others who made up the poor working-force of the nation. The term distinguished them from the impotent poor, who included the widows, orphans, the elderly and the infirm, as well as from the vagrants and the criminal elements on the fringes of society. The industrious poor were in no sense homogeneous in the manner of the nineteenth-century factory proletariat. They included a wide spectrum of occupations, and fine social gradations, as Dorothy George has shown.[45] Many were self-employed, and there was as much rivalry within this grouping as between it and the middling sort.

Finally, while it is anachronistic to write of class in the eighteenth century, it is often convenient to use modern terminology as long as it is recalled that it is merely a tool of analysis. Such generalisations have to be used if one is not to be buried beneath the weight of detail. Where necessary, qualifications will be made in the text and elsewhere.

[45] George, *London Life in the Eighteenth Century.*

PART ONE

1 The Provincial Hunger Riots of 1766

The hunger riots which spread across most of southern England in the summer and autumn of 1766 were the most extensive rural disorders in a century when food riots became chronic.[1] More serious in their threat to the social order than the violent protests against the high cost of 'necessaries' and the new Militia Act in 1756–7, the disturbances of 1766 placed a very heavy strain upon the forces of order. They foreshadowed the more serious agrarian riots of the next century.

While the War Office in September 1766 moved its detachments across the countryside in a vain effort to parry the rapidly-shifting threats from militant labourers, colliers, tinners, weavers, and others of the provincial dispossessed, the rioters became bolder and seized control of large tracts of the countryside almost in the manner of an occupying army. As the crisis developed, demands for military protection from market towns and isolated country estates poured into the War Office.[2] By late September the pattern of events had unfolded to the point where Lord Barrington, the Secretary-at-War, apprehended a threat of general insurrection. Striving to mobilise his limited resources, he ordered the commanders of both active troops and 'invalides' to assist the civil magistrates 'upon requisition',[3] while at the same

---

[1] Ashton, *Economic Fluctuations in England 1700–1800*, *passim*. The militia and food riots of 1756–7 were most serious in northern England when only about a quarter of the population lived north of the Trent (C. R. Fay, 'Significance of the Corn Laws in English History', *Economic History Review*, I, 1st ser. [1927–28] 314). In contrast the 1766 hunger riots affected seriously most of southern England, including some of the Midland counties where nearly three-quarters of the population lived.

[2] Barrington to the Earl of Suffolk, 1 October 1766, *Letter Book of Viscount Barrington*.

[3] Public Record Office, *Marching Orders of the Army*, WO5–54, p. 341 and *passim*.

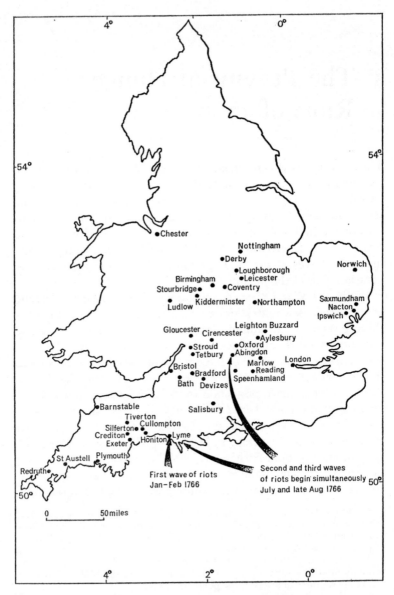

MAP 1   English hunger riots of 1766

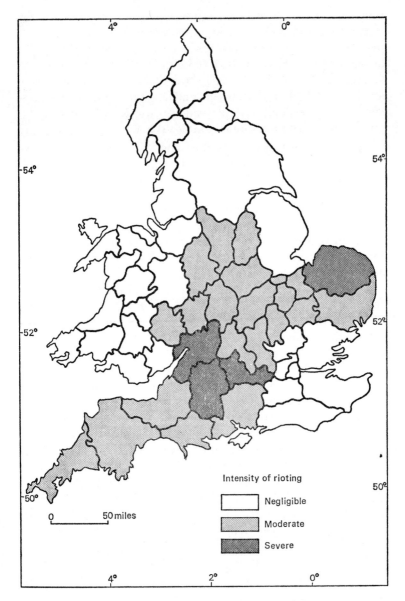

MAP 2   Disaffected counties in 1766

time he urged the leaders of rural society, the aristocracy and the gentry to abandon their lethargy and use their initiative in arming their servants to suppress less serious disturbances. On paper at least, the government had at its disposal sizeable forces. Sir George Savile calculated a mere two years later that there were available 18,000 regular troops and 33,000 militia to suppress general insurrection.[4] The personnel in the Royal Navy at this time amounted to more than 15,000. While many of these were in North America, it is noteworthy that a variety of naval ships were used to blockade the Thames during the riots of seamen and coalheavers in 1768.[5] In the agrarian disorders of 1766, the authorities relied on the army who only pacified the several disaffected counties with great difficulty. Although serious rioting had ended by late October, and trials before the special assizes began to relieve pressure on the crowded county jails by early December, hunger riots continued in a desultory fashion for the next two years in provincial England.

It will be valuable to examine first the timing, location, extent and direction of these riots to determine their immediate causes, before in subsequent chapters analysing the underlying factors behind the actions of both the rioters and the authorities.

In 1766 there were three waves of hunger riots. Occurring in January and February, the first wave essentially was a continuation of the disturbances of the previous year over high food prices and the construction of houses of industry in East Anglia. These riots were relatively minor, and except for sporadic outbreaks later in the spring they had ceased by the end of February. The second wave took place in the early weeks of August. This wave lasted only two weeks but, in the course of it, riotous mobs disrupted numerous districts in the West Country and Berkshire. The third, and most severe, wave of riots began in the first week of September and, except for minor isolated outbreaks, was over by the end of October. During this two-month period, much of southern England experienced serious disorders. The incidence of all three waves of rioting correlated closely with sudden fluctuations in the prices of grain and movements of wheat to the ports.

[4] West to Newcastle, 17 May 1768, Add. MSS., 32990.
[5] *Calendar of Home Office Papers* (1766–69) p. 371, no. 978; and Public Record Office, *Admiralty Entry Book* (1766–84) pp. 39–41. See also footnote 62, p. 36 below.

The riots of early 1766 were an expression of the unrest among the poor which had ebbed and flowed with seasonal and cyclical economic fluctuations during the course of several years. The prices of food had been generally high since the end of the Seven Years' War, due to disappointing harvests and epidemics among cattle and sheep. Discontent approached crisis proportions with the high cost of bread in 1765 and economic recession in parts of the north and the Midlands.[6] In that year, the bounty on grain exports had ceased when wheat prices at Bear Key, London, rose to 48/- a quarter.[7] But such self-regulating actions of the Corn Laws rarely solved the problem of high food prices in the eighteenth century. In less-accessible interior districts of England grain prices were frequently higher than they were at the ports where market prices determined export practices.[8] This situation led to serious tensions among the rural populace of inland regions. Disturbances at Braintree and elsewhere in 1765 caused Parliament to authorise the admission of duty-free grain between 10 May and 24 August.[9] As an additional cautionary measure Parliament granted authority to the government to suspend all grain exports during the summer recess of Parliament by Order-in-Council should circumstances warrant it.[10] (The granting of this power to the Ministry created confusion in the summer of 1766 when some politicians believed it was still in force.) Although some relief for the people came with the harvest of 1765, the prices of provisions remained high. By December one correspondent was forecasting general insurrection if prices did not drop and unemployment decline, especially in the northern industrial regions.[11] The riots which occurred in the following year however spread through southern England.

The earliest disturbances of 1766 took two forms: protests in the markets at the high prices of food and threatened attacks on flour mills, and renewed assaults on the recently-constructed workhouses in eastern England. As will be seen, the riots over the extension of a system of indoor relief frequently outside the recipient's parish of settlement were closely related to economic

[6] *Gentleman's Magazine*, xxxv (1765) 84–5 and 567.
[7] Ibid., p. 195.
[8] Ibid., p. 45.
[9] Ibid., p. 394.
[10] Harcourt to Jenkinson, 16 September 1766, Add. MSS., 38205.
[11] *Gentleman's Magazine*, xxxv (1765) 567.

deprivation, but they were somewhat more complex in origin than the common type of food riots. A considerable part of the rioters' hostility towards the workhouses related to resentment at the loss of traditional rights.

While the disturbances of early 1766 were relatively minor, they do offer some insights into the causes of unrest which later culminated in the serious outbreaks of violence in the summer and autumn of that year. They took place in and around the West Country grain port of Lyme. Here, on 24 January, the magistrates had to read the Riot Act to disperse 'one hundred ringleaders' and call for military assistance to deal with a mob of six hundred which the authorities expected to return.[12] Conditions remained threatening and the justices needed troops to maintain the peace,[13] until at the end of February the government took decisive action to remove the prime cause of unrest by introducing a parliamentary bill to suspend for six months all corn exports.[14]

The immediate occasion of the disturbances at Lyme had been the movement of grain through the port at a time when the usual seasonal increases were beginning to affect the prices of grain, which had been high since the autumn of 1765. Normally the greatest movement of grain prices occurred between Lady Day (25 March) and Michaelmas (29 September) when higher prices reflected dwindling stocks of old grain and the uncertainties of farmers and corn factors about the coming harvest.[15] Because Lyme was in the West Country, a region dependent upon other counties for grain to feed its people and stock, the local population was particularly sensitive to the collection of grain for export in times of scarcity and high prices. Probably in early 1766 there was an unusually large quantity of grain moving through the port. Because Europe and Britain tended to experience similar weather cycles, when harvests were poor in one country, they were poor across the continent and prices were universally high.[16] In 1766 famine conditions in Europe were draw-

---

[12] *Marching Orders*, WO5–54, p. 47.

[13] Ibid., p. 62.

[14] 6 Geo. III, caps. 3, 4, 5.

[15] See the records of grain prices on Lady Day and Michaelmas, 1752–64, of the bursar of Trinity College, Oxford, *Committee on High Prices of Provisions* (March 1765).

[16] C. R. Fay, *The Corn Laws and Social England* (Cambridge: Cambridge University Press, 1932) p. 28.

ing from England all the available grain. Probably shipments to the north of England, too, were unusually high. There had been considerable unrest in the industrial north, and the government, fearing outright rebellion, had already begun to reinforce its northern garrisons.[17] With the memory of the 1756–7 hunger riots fresh in mind, the Ministry understandably was anxious to ensure an adequate supply of bread for the northern populace. Thus it encouraged shipments of grain through the western ports to supply the northern population.

The government's action in passing through Parliament legislation which suspended the exportation of grain between 26 February and 26 August clearly illustrates the importance of examining the role of the authorities in the agrarian disturbances of the 1760s. This suspension of corn exports was the vital move which averted further riots in the grain ports of the West Country and the depressed industrial areas of the interior of England. In acting thus, the Ministry was remarkably provident. Politicians were usually reluctant to interfere with corn regulations, which they supposed to be in the interests of the landed classes and the general population. Early in the previous decade Henry Pelham, addressing the gentry, freeholders and clergy of Sussex, had typically affirmed his government's support for the corn laws.[18] Ministers were unwilling to antagonise grain interests or to jeopardise European grain markets, and they frequently gambled on bountiful harvests to relieve grain shortages. Usually embargoes on exports were too late to prevent serious food shortages. When such grain suspensions were imposed for several months in a year, they tended to result in the concentration of grain exports within a limited period, rather than to reduce the total quantity exported in a given year.[19] The consequence of this was that the

[17] *Marching Orders*, WO5–54, p. 182.

[18] Add. MSS., 32732, fol. 570, cited in Donald Grove Barnes, *A History of the English Corn Laws from 1660–1846* (New York: Augustus M. Kelley, 1961) p. 46. See also Harris to Hardwicke, 3 October 1766, Add. MSS., 35607, fol. 295; and Newcastle to White, 17 November 1766, Add. MSS., 32977, fol. 403–4.

[19] Contrast the arguments of R. B. Rose, who suggests that popular riots were successful in mitigating the worst effects of the government's protectionist policies. R. B. Rose, 'Eighteenth Century Price Riots and Public Policy in England', *International Review of Social History*, VI, no. 2 (1961) 277–92.

accelerated rate of export incited further popular unrest. This is what occurred in the summer of 1766. But in February 1766, by obtaining an embargo upon corn exports, the Ministry was responding to more than the mere threat of food riots over grain movements in the West Country, which as yet scarcely warranted the 'narrow-bottomed' Rockingham Ministry's adding the grain lobby to the other commercial and industrial interests already agitating against its commercial policies towards America.[20]

In fact the government was responding to political as well as economic and social circumstances. In late February the question of the repeal of the Stamp Act was causing unrest and uncertainty among various interests in the country. The Commons had only agreed on repeal in the fourth week of February after a bitter struggle, and there was a real prospect of an even tougher fight in the Lords. Horace Walpole's comments are illuminating here :

A general insurrection was apprehended as the immediate consequence of upholding the bill, the revolt of America, and the destruction of trade was the prospect in the future. A nod from the ministers would have let loose all the manufacturers of Bristol, Liverpool, Manchester, and such populous and discontented towns, who threatened to send hosts to Westminster to back their demands.[21]

Allowing for Walpole's *penchant* for exaggeration, one may recognise that there was a great deal of unrest among the industrial workers at this time, which formed the background to the government's decision to seek a suspension of the grain exports for six months. The effect of this suspension was to emasculate any potential rebellion by removing the worst resentments of the unemployed manufacturers, who normally produced goods for the American market. Lowering the prices of food temporarily appeased the poor, while the opportunity to import duty-free corn pleased some grain merchants and their American suppliers. Because the legislation suspending grain exports came into effect

---

[20] John Brooke, *The Chatham Administration, 1766–1768* (London : Macmillan, 1956) p. xi *et seq.*

[21] Horace Walpole, *Memoirs of the Reign of King George the Third*, re-ed. G. F. Russell-Barker (London : Lawrence & Bullen, 1894), II, 211–12; see also Ian R. Christie, *Crisis of Empire, Great Britain and the American Colonies, 1754–1783*, 'Foundations of Modern History', ed. A. Goodwin (London : Arnold, 1966) pp. 60–1 and *passim*.

three weeks before the Lords agreed to the repeal of the Stamp Act, it provided interim relief before the ending of the American non-importation agreements increased employment in the depressed regions.

Whatever the motives of the government in safeguarding the supplies of grain in the country, their action in ending the provocative corn movements to the western and northern ports was timely, and most of the provincial population remained calm until prolonged bad weather and resumed grain movements to the ports in anticipation of the lifting of the export ban threatened to create famine conditions.

For the next five months, between, that is, February and August, the poor remained relatively calm. The few minor challenges to public order came from sporadic smuggling riots and renewed attacks on Suffolk workhouses. Both types of disorders were related to unemployment and high food prices. Plainly, factors other than economic deprivation help to account for the remarkable increase in 'owling'[22] and smuggling in the 1750s after a period of decline attributed to harsh punishments by some contemporaries.[23] Certainly the tariff policies of the government in the second half of the century and fluctuations in European economies contributed to the expansion of this illicit trade. Yet men plagued by rising prices and insufficient wages readily sought extra income. Smuggling offered a convenient source of income to many seamen beached when the prosperous war years ended, and to the poor of coastal regions who felt the pinch of adversity after 1764. It is not coincidental that in the second half of the century when food riots were commonplace, smuggling became such a large industry. The connection between economic unrest and attacks on East Anglian workhouses is similarly direct. The building of the new 'houses of industry' to provide the indigent with indoor relief in an institution which served several parishes rather than outdoor relief in their parishes of settlement, was an effort by the ruling orders to reduce the cost of welfare in a period of general economic difficulty. This deprivation of what the poor regarded as their 'freehold' incensed them at a time when the price of 'necessaries' threatened to force many of the

---

[22] Illicit exports, usually of wool.
[23] *Gentleman's Magazine*, xxvii (1757) 528.

marginally indigent permanently on to poor relief.[24] While the loss of one of their ancient liberties was pre-eminently of concern, the poor readily recognised the relevance of high food prices to their difficulties. One witness to the destruction of a partially-built workhouse at Bulcamp, Suffolk, reported the rioters' declaration that as they had had success 'in this their first undertaking, they would reduce the price of corn or pull down all the mills about'.[25]

But the attacks on poor law institutions in 1766 were not as serious as they had been in the previous year, when for a while the government contemplated establishing a special commission to make severe examples of the rioters.[26] In general, the government's policies of permitting duty-free imports of grain, suspending grain exports and repealing the Stamp Act reduced tension in the provinces by lowering the prices of provisions and providing work for many of the industrial workers. Probably this relaxation of tension averted insurrection, for example in the hardware-producing towns of Birmingham, Walsall, Wolverhampton, and Sheffield, which in February were reportedly in great distress due to the decline of trade with America.[27] Certainly such commercial centres as London, Bristol and Liverpool responded favourably to the government's actions. Although many of the old cloth centres of the south of England continued to feel the effect of West Riding competition in worsted cloth production, the resentment of the poor seems to have died down with the ending of provocative grain shipments abroad after 26 February, and they were content to await the harvest in expectation of much lower prices.

Evidently reassured by this absence of serious rioting between February and August, the authorities also banked upon the lowering of prices to acceptable levels in the autumn as a result of a plentiful harvest, and increased opportunities for employment for the industrious poor with renewed American trade. They did not

---

[24] Several thousand rioters destroyed a newly-built workhouse near Saxmundham in Suffolk, and several were killed by the military. *Gentleman's Magazine*, xxxv (1765) 392. Four hundred rioters defied the magistrates at Nacton. *State Papers*, PRO/SP 37/4, fol. 202/595.

[25] *State Papers*, PRO/SP 37/4, fol. 196/595.

[26] *Marching Orders*, WO5–54, pp. 53–4, 58, 246; and Public Record Office, *Domestic Entry Book*, vol. 25, pp. 160–4.

[27] *Annual Register*, ix (1766) 61.

therefore contemplate an extension of the embargo on grain exports beyond 26 August before Parliament recessed for the summer. But there were already in early July indications that their gamble was a long shot at best. A late frost shortly after seed-time had damaged the young seed in the ground, and heavy flooding from the prolonged July rains had compounded the farmers' problems.[28] The normal results of such wet growing conditions is a tendency to produce swollen ears of wheat, which yield a coarse, lightweight grain when threshed. Although the coarser grains like barley and oats were not as vulnerable to harsh weather conditions as wheat, excessive rain encouraged the growth of weeds which adversely affected all crops. While conditions varied from district to district, several of the counties which later were seriously affected by the hunger riots reported the effects of prolonged bad weather in the early summer. In describing the floods in Gloucestershire, Oxfordshire and 'adjacent counties', especially Worcestershire, one newspaper claimed that it was the wettest summer since 1733.[29] Both grain crops and livestock were adversely affected by the bad weather. Oxfordshire reported much spoilt hay on the low ground, and some 3000 sheep were lost around Wisbech.[30] At Maidenhead and elsewhere in Berkshire, floods reportedly covered many Thames-side fields and heavily damaged the hay.[31] In counties such as Wiltshire many sheep drowned.[32] One measure of this natural disaster was an eight per cent increase in the price of wool within a fortnight.[33] Gloucestershire and Wiltshire reported that heavy rains had damaged both crops and livestock.[34]

Inevitably the immediate effect of such natural disasters was to push up the prices of meat, wool and bread, and thereby to increase the privations of the poor.[35] Ironically the first incidents in the second wave of hunger riots in 1766 began the very day

[28] *Public Advertiser*, 10, 27, 30 July 1766. *Gazetteer and New Daily Advertiser*, 10, 12, 26, 30 July 1766.
[29] *Gazetteer and New Daily Advertiser*, 26 July 1766.
[30] *Public Advertiser*, 4 August 1766.
[31] Ibid., 12 July 1766.
[32] Ibid., 10 July 1766.
[33] *Gazetteer and New Daily Advertiser*, 14 July 1766.
[34] *Public Advertiser*, 10 July 1766.
[35] *Committee on the High Prices of Provisions* (March 1765).

that an Order-in-Council called for public prayers for the end of the rains.[36]

This wave of riots extended through the West Country and the county of Berkshire between 30 July and 12 August. Stoke, Sidbury, Ottery St Mary, Crediton, Honiton, Exeter and Lyme in the West Country were the first to witness the destruction of mills by rioting mobs of up to five hundred persons.[37] Newbury, Shaw and Speenhamland in the home county of Berkshire experienced similar outbursts, in which the mobs attacked and robbed the premises of mealmen especially.[38] The final disorders of this wave took place a few days later. Again they were in the west, at Usseolm, Lemnion, Cullompton, Bradnick, Tiverton, Silferton and Barnstaple.[39]

Conceivably the psychologically depressing effect of such a sustained heavy rainfall was considerable and may partly account for the outburst of rioting at this time. But the choice of targets for their frustrations by the rioters suggests that these were social protests rather than the expression of blind resentment against the elements.

The direction of these riots was against bunting mills where flour was dressed, starch mills which catered to the conspicuous consumption of the privileged, and the premises of mealmen and bakers.[40] Often rioters destroyed property, including stocks of food, which suggests anger and frustration rather than outright starvation. Most frequently the protesters imposed the sale of provisions at what they felt to be 'just' prices.[41] Probably the bakers bore the brunt of the rioters' attacks. With some justification the poor blamed bakers for the high cost of bread. For their part the bakers attributed the problem of high prices to a brewers' monopoly on yeast, and claimed that the law closely regulated bakers and prescribed their profits on the basis of the lower prices prevalent forty years earlier.[42] Probably the bakers' claims would have been nearer the truth before 1758 than after.

[36] *Gazetteer and New Daily Advertiser*, 30 July 1766.
[37] Ibid., 6 August 1766. *Annual Register*, IX (1766) 124.
[38] *Gazetteer and New Daily Advertiser*, 9 August 1766.
[39] *Public Advertiser*, 12, 14 August 1766.
[40] *Marching Orders*, WO5–54, p. 270.
[41] Rose, 'Eighteenth Century Price Riots', pp. 277–92.
[42] *Gazetteer and New Daily Advertiser*, 10 July 1766.

In that year a new act, by permitting either variation in the weight of loaves while the price remained fixed, or variation in the price of bread while the weight of loaves remained constant, 'threw the whole system of bread regulation into confusion' and enabled bakers and grain dealers to make large profits at the expense of the poor. Where the setting of the bread assize had been abandoned, that is, in many rural regions, the bakers were freer still to exploit their positions.[43] Frequently bakers found ways of defeating its intent, even where magistrates set the assize of bread regularly. One newspaper, for example, reported the purchase of a single load of grain 'by which, and not by the current market price, the bakers regulate their assize of bread'.[44]

Millers shared with bakers the immediate resentments of the poor in times of scarcity. The populace with good reason suspected many millers of dishonesty, and feelings ran very high when rioters discovered stocks of chalk, alum, whiting, pease meal and other adulterants. The chief reason for singling out bunting mills, however, was almost certainly that millowners, many of whom had developed into large-scale grain merchants by this period, were forcing up the prices of grain and flour by engrossing supplies in anticipation of a poor harvest and the resumption of grain exports after 26 August.[45] Already in early August there were reports of a rapid rise in the prices of grain.[46] The normal response of prices a few weeks before the end of a grain export embargo was to rise. In such circumstances in 1758, a correspondent of the *Gentleman's Magazine*, commenting upon the artificially high cost of grain, noted : 'And yet how is the price kept up beyond every man's expectation? Why truly our great growers thrash out little or none; for, say they, after Christmas the ports will be opened for exportation, and the distillers permitted to distil from grain again.'[47] In July 1766 the prices rose higher than ever because the wet weather depressed popular expectations of a normal harvest, and confirmed the wisdom of farmers and dealers who were holding on to their stocks in expectations of great profits.

---

[43] Sydney and Beatrice Webb, 'The Assize of Bread', *Economic Journal*, xiv (June 1904) 196.
[44] *Public Advertiser*, 22 August 1766.
[45] 'A Letter from Portsmouth', ibid. 15 August 1766.
[46] *Gazetteer and New Daily Advertiser*, 2 August 1766.
[47] *Gentleman's Magazine*, xxviii (1758) 509.

The proximity of the initial outbreaks to the western grain port of Lyme, the scene of earlier disturbances before the embargo was imposed, suggests renewed popular concern at grain movements causing high prices and threatening outright famine. Newspaper reports that the mob at Usseolm and Lemnion blamed exports for their distress corroborates this.[48] The only legal 'exports' of grain at this time were shipments to the north of England, and these were probably what the poor resented. It is possible however that in the West Country, where smuggling was highly organised, some factors were exporting grain to famine-stricken Europe at this time. In September the War Office was obliged to order commanders at Canterbury, Lewes and Padstow to transfer detachments to patrol the Suffolk coast to halt the evasion of the embargo on grain exports.[49]

Like the West Country counties, Berkshire was not a heavy grain producer but much grain passed through it on the way to a port of shipment. Always in times of crisis Berkshire felt the draining power of London. The upper Thames valley provided a readily accessible route to the leading grain exporting port in the country. Not surprisingly, in July 1766 riots broke out in Berkshire at the same time that Lyme, the West Country grain port, experienced disorders.

The location of the initial outbreaks in the second wave of rioting in Berkshire and at Lyme, then, indicates the provocative nature of grain exports in time of expected famine and the impact of government policies on events. As early as one month before the ending of the six-month prohibition on grain exports, correspondents of the *Gentleman's Magazine* asked the Privy Council to prevent the loading of grain ships, and these demands in the press and elsewhere became more strident until the Privy Council belatedly imposed a new embargo on 26 September.[50]

There was now another, but much shorter, lull after the second wave of riots ended on 12 August. It lasted scarcely three weeks. During this period more optimistic forecasts of the harvest, encouraged by the dry August weather, brought down prices and lessened public concern that continued grain movements through

---

[48] *Public Advertiser*, 12 August 1766.
[49] *Marching Orders*, WO5–54, p. 353.
[50] *Gentleman's Magazine*, xxxvi (1766) 389.

the countryside to the ports would cause outright famine. Speculators released grain in expectation of lower prices resulting from a good harvest and were reluctant to buy more to maintain their stocks.[51] The remedial actions of the gentry and the magistrates in making available cheaper grain and flour reassured the poor also. At Exeter magistrates fixed wheat prices at 5/6 per bushel, although the farmers reportedly wanted 8/- or 9/-.[52] Elsewhere, gentlemen bought flour and sold it to the poor at 3½d per pound.[53] All these factors resulted in a spectacular drop in grain and flour prices. A letter from Shrewsbury dated 20 August reported wheat prices down to 6/- a bushel from a high in Wales of 11/-.[54] Yet discontent stayed close to the surface, and troops remained in many areas. Bakers in Newbury, Berkshire, who had ceased their sales of bread early in August, did not resume them until 27 August, when a soldier was placed at every bake shop door.[55] It is quite possible that minor outbreaks continued unreported in the metropolitan newspapers, until in early September the final, and most serious, wave of food riots swept across southern England.

The placing of the initial riots of this wave at the beginning of September is important because it brings out clearly the effect of the Ministry's failure to renew the embargo on grain exports when it expired on 26 August, and raises questions about the government's responsibility for the chain of events which followed. George Rudé writes of a six weeks' break in rioting in the summer of 1766, although he does not entirely discount the possibility of riots occurring in this period unreported in the press.[56] He places the resumption of rioting on 23 September, and bases this assertion on reports in the *Annual Register* and the *Gentleman's Magazine*. He cites, however, in his footnotes Treasury Solicitor's papers relating to the Berkshire session of the Special Commission in which prisoners are listed as participating in riots as early as 6 September.[57] London newspapers and the *Marching Orders of the Army* indicate the resumption of rioting several

[51] *Public Advertiser*, 14, 15 August 1766.
[52] Ibid., 12 August 1766.
[53] *Gazetteer and New Daily Advertiser*, 6 August 1766.
[54] *Public Advertiser*, 25 August 1766.
[55] Ibid., 27 August 1766.
[56] Rudé, *The Crowd in History*, p. 41.
[57] Public Record Office, *Treasury Solicitor's Papers*, T.S. 11/995/3707.

days before this date. In fact, by 23 September the riots had
gathered alarming momentum and the government's concern is
evident in the instructions sent from the War Office to commanding officers.[58]

This coincidence of renewed grain exports and popular protests
indicated the alarm of the poor at the prospect of outright famine.
John Pitt, attorney and steward of Lord Hardwicke's Gloucestershire estate, substantiated this view when he noted that the chief
direction of the riots in the West Country was against exports
and the attacks were mainly on 'general reservoirs and the sale
of flour'.[59] A critic of Arthur Young's *Six Weeks' Tour Through
the Southern Counties of England* reached the same conclusion.
Writing in the *Lloyd's Evening Post* in 1768, he condemned the
government's failure to close the ports at the end of the Parliamentary Session in July 1766, which 'drove those unhappy persons to that dreadful alternative of either starving or hanging'.[60]
In similar terms one Norfolk correspondent wrote of the 'terror
of the poor' in and around the metropolis at the export of grain.[61]
Some businessmen even found it necessary to declare publicly
that they had not engrossed provisions or exported 'wheat, flour,
or any other grain'.[62] Apparently a sudden fear of impending
famine brought on by large-scale grain movements swept through
the ranks of the poor, causing them to protest violently. This fear
was only matched later by the alarm of the aristocracy and gentry
on their isolated estates at the extent of the disorders and the
prospect of revenge at the hands of the dispossessed.[63]

The most disaffected counties now were Berkshire, Gloucestershire, Norfolk and Wiltshire, to judge from the records of the
Special Commission later appointed to empty the crowded county
jails and dispense swift retribution to the prisoners, who were too
numerous for the regular assize courts to handle. But riots affected

[58] *Marching Orders*, WO5–54, *passim*. See also *Calendar of Home Office
Papers* (1766–9) p. 80, no. 373–4; and *Domestic Entry Book*, vol. 142,
pp. 4 and 9–11.
[59] John Pitt to Hardwicke, 19 December 1766, Add. MSS., 35607, fol.
339.
[60] *Lloyd's Evening Post*, 25–27 May 1768.
[61] *Public Advertiser*, 9 September 1766.
[62] James Townsend, ed., *News of a Country Town* (London: Humphrey
Milford, 1914) p. 58.
[63] Compare Georges Levebvre, *La Grande Peur* (Paris: Société d'Edition
d'Enseignement Supérieur, 1956).

most counties in the Midlands, in the West Country and around London. Nottinghamshire, Derbyshire, Worcestershire, Northamptonshire, Leicestershire, Bedfordshire, Hertfordshire, Oxfordshire, Buckinghamshire, Suffolk, Hampshire, Somersetshire, Devonshire and Cornwall were the most important of these. Urban centres like Gloucester, Bristol, Nottingham, Derby, Birmingham and Norwich had particularly serious disorders.

London, the south-eastern counties of Kent and Surrey and the northern counties, except for minor, isolated incidents at Whitehaven, Carlisle and Berwick which were usually related to grain exports, remained quiet. Because the conditions of the metropolis were unique, an explanation of its freedom from food riots in 1766 is offered elsewhere. There were several reasons for the calm of the populace in the north and the south-east. In most of these districts the magistrates were energetic in controlling sales by sample and ensuring that adequate supplies of grain were available. Thus Durham justices insisted on farmers offering all their grain for sale in the open market.[64] The result of such actions was both psychological and material. First, the revelation of the actual prices of commodities calmed the suspicions of the poor that bakers and others were making enormous profits at their expense, something not apparent when markets were by-passed by farmers and dealers; second, the provisions' prices did not rise rapidly. Surrey markets, for example, reportedly sold wheat at 5/- a bushel owing to the 'vigilance of the magistrates'.[65] In the north, too, lower prices were due to more favourable harvests than in the south. Cumberland for example claimed the best hay and corn harvest in living memory;[66] while crops in Yorkshire were generally heavy.[67] Farmers in the market at Bishop Auckland, County Durham, were able to sell their wheat at 4/- per quarter.[68] Apart from the lower costs of grain, the willingness of northerners to avail themselves of other coarser grains and foods which the southerners regarded with distaste meant that no serious shortage of food was experienced in the north. Labourers and poor manufacturers of the north customarily ate potatoes and oatmeal and

[64] *Public Advertiser*, 7 August 1766.
[65] *Gazetteer and New Daily Advertiser*, 6 November 1766.
[66] Ibid., 7 September 1766.
[67] *Public Advertiser*, 4 August 1766.
[68] Ibid., 7 November 1766.

were therefore less affected by the cost of wheat and meat.[69] The income, too, of many northern workers suffered less in the recession of 1766 than was the case for industrial workers in the Midlands and southern England. While it is true that pitmen in the north-east of England were restless in 1765 and 1766, their industry was not depressed in the way that for example the Midland hardware industry was. Frequently where northern workers lost their jobs, they could find work in the expanding woollen-worsted industry of the West Riding.[70] In contrast, the cloth centres of southern England experienced severe trade fluctuations, and unemployment was high among weavers, combers, and others of the cloth industry.[71] Yet the unrest of the Newcastle and Sunderland pitmen had alerted the government to the dangers of insurrection. Mindful of the extensive riots in the north over food prices and the new Militia Act in 1756–7, the authorities took care in 1766 to ensure an adequate supply of food, while at the same time they reinforced their garrisons in Newcastle and elsewhere to meet all eventualities.[72] No such provisions were made in southern England. The foresight of the authorities helped to avert disorders in the north in the summer of 1766. But probably the most important reason for the absence of serious riots in the northern counties was the fact that relatively little grain moved through northern ports in that year. Six-sevenths of all grain which received bounty payments in 1766 went through the port of London.[73] How much of the other one-seventh went through the western grain ports is not clear, but certainly it was a sizeable part. Thus the provocation of grain movements to the northern ports in a period of relative scarcity was absent.

The initial pattern of disturbances in the third wave of rioting

---

[69] See Fay, *The Corn Laws*, p. 4, citing Charles Smith, 'Three Tracts on the Corn Laws and Trade', 1766 ed., Supplement, p. 182.

[70] Herbert Heaton, *The Yorkshire Woollen and Worsted Industries: From the Earliest Times up to the Industrial Revolution*, vol. x of *Oxford Historical and Literary Studies* (London: Oxford University Press, 1965) p. 279.

[71] Ephraim Lipson, *The History of the Woollen and Worsted Industries*, Histories of English Industries, ed. E. Lipson (London: A. & C. Black, 1921) p. 253.

[72] *Marching Orders*, WO5–54, p. 182.

[73] An account of the total sums paid for bounties, 1766–81, William L. Clement Library, Ann Arbor, Michigan, *Shelburne Papers*, vol. 135.

in 1766 confirmed the direct relationship between grain movements and disorders. As in the second wave, the first outbreaks
occurred almost simultaneously in the West Country and Berkshire, regions through which grain for export passed. Military
reports noted the disturbance of the markets at Cullompton and
Ottery St Mary (Devon) and the destruction of most bolting mills
within a vicinity of twenty miles by 'dangerous and riotous'
mobs.[74] A day or two after, on 6 September, two Berkshire country towns were disturbed. At Abingdon 'many riotous persons'
led by a bargeman took grain from farmers and distributed it;
while at Drayton a crowd of labourers stole wheat, flour and
other provisions.[75] Later, riots became much more extensive in
the West Country and Berkshire.

After these initial outbreaks, no clear pattern of expansion
from the two areas is apparent.[76] Riots occurred throughout the
southern counties more or less spontaneously. The timing of these
disturbances probably reflected differing harvest and threshing
times, the impact of market-borne rumours, the contagion of
general disorder in the countryside, heavy buying by London
dealers or specific local grievances such as the construction of
houses of industry in East Anglia.

Such outbreaks were too numerous to catalogue here; but it
will be useful to note when each county was first affected and
how long rioting continued. Within a few days of the initial
outbreaks in the West Country, 'regulators' were at work in the
Midland markets of Stourbridge (Worcestershire), Birmingham
(Warwickshire) and Whitney (Gloucestershire), forcing down the
prices of provisions.[77] Shortly thereafter rioters disrupted Birmingham Fair and rescued prisoners from jail.[78] In the second week
in September, mobs began sustained attacks on houses and mills,
destroying furniture and removing food, in several towns and
parishes of Gloucestershire, a county in which all markets were
reportedly under the influence of the mob by the end of Septem-

---

[74] *Marching Orders*, WO5–54, pp. 294–5.
[75] *Treasury Solicitor's Papers*, T.S. 11/995/3707.
[76] Contrast Rudé's somewhat oversimplified map illustrating the spread of
food riots in 1766. Rudé, *The Crowd in History*, p. 40.
[77] *Gazetteer and New Daily Advertiser*, 8, 10, 12 September 1766.
[78] *Marching Orders*, WO5–54, pp. 347–8.

ber.[79] Throughout most of September and October, large crowds repeatedly disturbed Stroud, Gloucester, Cirencester and Tetbury.[80] By the third week in September riots were widespread in Wiltshire. On 19 September several parishes of Bradford (Wiltshire) saw a great number of 'idle and disorderly persons assembled in a riotous manner' attack homes and mills, destroy furniture and windows, and steal bacon and other provisions.[81] Very quickly the rest of the county became involved. Devizes and Bradford continued to be centres of revolt until the end of October, and required the stationing of troops for several weeks.[82] On 20 September rioters destroyed mills at Tiverton (Devon), and combers, labourers and weavers sent a 'threatening and incendiary letter' to the corporation. As late as 11 November the unsettled situation required the presence of a company of soldiers.[83] Reports of further disturbances came from Cornwall, Devon and Bristol on 19 September.[84]

This final wave of hunger riots, which spread in an irregular pattern across southern England, reached its peak in the fourth week of September. The War Office, now thoroughly alarmed, sent a flood of orders to commanding officers to aid the magistrates fully 'upon requisition'. Lord Barrington ordered the dragoons, who had hitherto been forced to march dismounted to troubled regions, to 'take up their horses from grass' in distant pastures.[85] The Minister-at-War alerted not only twenty-two regiments of foot and dragoons, but also fourteen independent companies of 'invalides' in various centres as far west as Plymouth and as far north as Newcastle.[86] The third phase of this last wave of riots in 1766 now began.

The provincial ruling orders, after their initial lethargy, flooded

---

[79] Pitt to Hardwicke, 29 September 1766, Add. MSS., 35607, fol. 291.
[80] *Treasury Solicitor's Papers*, T.S. 11/5956/Bx 1128.
[81] Ibid., T.S. 11/1116/5728.
[82] *Marching Orders*, WO5–54, pp. 341, 357.
[83] Ibid., pp. 308–9.
[84] A letter from Wiltshire, dated 20 September, *Gazetteer and New Daily Advertiser*, 26 September 1766.
[85] Instructions to ten dragoon regiments at Colchester, Manchester, Lewes, Blandford, Worcester, Coventry, Northampton, Leeds, Stamford and Derby, dated 23 September 1766, *Marching Orders*, WO5–54, p. 315. All regiments to take up horses from grass in south Britain to assist justices in riots, *Calendar of Home Office Papers* (1766–9) no. 277, 25 September 1766.
[86] *Marching Orders*, WO5–54, pp. 318–20.

the War Office with demands for military assistance. But the riots had already gained momentum and they continued to spread during the next few weeks. By 27 September mobs were active in Hertfordshire forcing down the prices of food.[87] About the same time the destruction of mills and other disturbances obliged the drafting of dragoons into Norwich.[88] In early October the War Office had to send detachments into Leicester, where 'numerous and disorderly persons have assembled . . . and committed great acts of violence and outrage'.[89] Disorders affected Coventry at about the same time.[90] Later in October a troop of dragoons had to suppress a riot in the vicinity of Loughborough.[91] On 4 October Nottingham, Oxford, Leighton Buzzard (Bedfordshire) and Great Marlow (Buckinghamshire) reported serious riots.[92] Several days later Derby was similarly affected.[93] Ipswich, the scene of a number of attacks on the new houses of industry earlier in the year as well as in 1765, continued to be the centre of insurrection in October. Success in their attacks on poor-law institutions in East Anglia had encouraged the rioters to attempt to lower food prices too. On 20 October they seized butter and sold it at lower prices than the farmers asked and threatened to burn the town.[94]

Gradually the combined efforts of the magistrates and the army proved successful and most of the disaffected regions were quiet by late October. The four most seriously disturbed counties, Berkshire, Gloucestershire, Wiltshire and Norfolk, ceased to report major riots after mid-October, although the authorities built a wall in Gloucester market to protect the soldiers, and commanders continued to detach companies through the country districts to assist the magistrates.[95] Elsewhere riots were over by the end of October, although isolated incidents at Bristol, Ludlow, Chelmsford and Birmingham continued until late November. Upon the return of relative calm, the local authorities with the

[87] *Public Advertiser*, 27 September 1766.
[88] *Marching Orders*, WO5–54, p. 326.
[89] Ibid., p. 337.
[90] Ibid., p. 342.
[91] Ibid., pp. 367–8.
[92] *Gazetteer and New Daily Advertiser*, 4 October 1766.
[93] Ibid., 13 October 1766.
[94] *Marching Orders*, WO5–54, p. 365.
[95] Ibid., p. 356.

aid of troops devoted their energies to hunting down the ring-leaders and filling the county jails, which remained overcrowded until the hearings of the Special Assizes in December 1766.

The timing of the last wave of hunger riots in the summer and autumn of 1766 and their direction indicate the importance not only of the resumed grain exports but also of fluctuating food prices as a precipitating cause of popular disorders. There were three general reasons why an extraordinary price rise occurred after the end of August. First, there were very poor harvests across most of Europe in 1766.[96] Russia, Turkey, France, Spain, Portugal and Italy were particularly adversely affected.[97] Germany, Holland, England and the Scandinavian countries were called upon to supply the deficit from their own stocks.[98] Lord Shelburne's commercial correspondents in Amsterdam, David Barclay and Sons, wrote in early October of Holland being denuded of grain due to immense shipments to Italy.[99] In England, factors and agents received from Europe commissions to buy at an 'unlimited price'.[100] Second, adverse weather conditions in England resulted in a crop two-thirds the usual size and inferior in quality.[101] Third, an export embargo on grain had prevented the supply of England's traditional markets in Europe between 26 February and 26 August, during which time a pent-up demand for English grain developed. Even before the lifting of the embargo, grain ships were loading in London and the outports. After 26 August, grain poured out of the country at an alarming rate, until the Privy Council reluctantly proclaimed a further prohibition on exports on 26 September.[102] Thus most of the grain exports of 1766 took place within the space of one month, the other two months of free exports being in the depth of winter when commerce was normally greatly curtailed by poor

[96] Horace Walpole complained of the excessive quantity of grain exported. Horace Walpole to Sir Horace Mann, 25 September 1766, Horace Walpole, *Letters*, ed. Paget Toynbee, 19 vols (Oxford: Clarendon Press, 1903–25), VII, 42.

[97] *Gazetteer and New Daily Advertiser*, 11 September 1766, reported south Italy except Sicily with a great grain shortage and prices double normal. See also *Public Advertiser*, 27 September 1766.

[98] *Gazetteer and New Daily Advertiser*, 17 September 1766.

[99] *Shelburne Papers*, vol. 132, fol. 65.

[100] Ibid., fol. 19–20.

[101] Ibid., fol. 65.

[102] *Gentleman's Magazine*, XXXVI (1766) 399.

weather. Grain movements in September were therefore very evident and not only helped to force up home prices but incensed the population. The reaction of Norwich rioters was typical. On 28 September 1766 they destroyed malt which had been 'entered with the proper officer of excise'.[103] This concentration of grain exports within the space of one month was serious enough, but its association with the ending of American grain imports had an immediate effect on prices and created a very serious situation for the government. As a gentleman returning from a tour of the western counties observed: '. . . the moment advice came of importation being stopped, and exportation allowed, the great farmers and corn dealers began to combine and the poor to murmur, which is one cause of the riots.'[104]

Price structures in local markets differed in the speed of their response to the restoration of exportation, and the prohibition of the free importation of American grains. The first big rise came in the west where newspaper correspondents reported 'monopolising farmers' were buying up grain stocks.[105] After a mere three days of exports, journalists estimated the average of the western markets was 7/- a bushel.[106] Less than two weeks later the price of bread in London had reportedly risen half an assize,[107] while the buying of grain for export around Carlisle had raised the price per bushel by 2/-.[108] Plainly, with the great demand for grain in London both for export and consumption, grain dealers now moved out in an ever-widening radius from the capital, buying up grain and forcing up local prices in the process.

Prices rose most steeply in areas where threshing had already revealed the poor quality of local crops. In districts where low wages were current, employment uncertain or payments in kind made in lieu of money wages, that is, in the old cloth counties of the west and East Anglia, in the hardware or coal-producing Midlands and elsewhere, the hardships of the poor were most severe. When these districts were not themselves major grain producers, but merely corridors along which great quantities of

[103] Norwich Record Office, Norwich, Deposition of John Glover, City Merchant, *Depositions and Case Papers*.
[104] *Gazetteer and New Daily Advertiser*, 8 September 1766.
[105] Ibid., 28 August 1766.
[106] Ibid., 29 August 1766.
[107] Ibid., 5 September 1766.
[108] Ibid.

grain moved to London or the outports, the reaction of the populace was most immediate and violent.

Berkshire to the west of London provided such a corridor between the great corn counties of Norfolk, Lincolnshire, Suffolk, Cambridge, Rutland, Hertfordshire, Bedfordshire and Buckinghamshire on the one hand, and either London or the western ports on the other. Although not itself a significant grain-growing county, Berkshire always felt the draining power of the 'Great Wen' down the upper Thames valley in times of scarcity, and prices in local markets rose quickly. By early September 1766 it was apparent that harvests had been especially poor in the corn counties, particularly in Norfolk and Suffolk,[109] and there were reports of several thousands of quarters of grain from the western counties brought down the Thames in barges, as London dealers sought to make good the deficiencies of their East Anglian suppliers.[110] Such grain movements were a common link between the disorders of the West Country and Berkshire.

Price rises which had first occurred in the grain ports in early September quickly spread to other regions as threshing revealed the low quality of the grain crop, 'badgers' buying up supplies for the urban centres pushed up local prices, or rumours of famine spread through the markets. Frequently the actions of local authorities and private gentlemen in offering grain to the poor at low prices mitigated the worst effects of spiralling prices for a while. Thus distance from London or the outports did not solely determine price fluctuations.

One important feature of the price rise in August and September was that it was the latest of several price fluctuations which occurred over a period of several months and which composed a phase in a general increase in food prices which had begun in 1764. These price movements followed a six-year period when the real income of the industrious poor made a significant advance, which T. S. Ashton ranked with that of the 1730s.[111] In 1766 the severe price fluctuations were directly affected by the expectations of the poor about the available food supply.

The universal concern of eighteenth-century Englishmen with harvest prospects is readily apparent from the space devoted to

109 *Public Advertiser*, 10 September 1766.
110 *Gazetteer and New Daily Advertiser*, 1 September 1766.
111 Ashton, *Economic Fluctuations*, p. 22.

weather reports and the state of the crops in contemporary journals. In years of poor weather, concern increased as the harvest
time approached and prices rose. July, August and September
were frequently months when magistrates braced themselves for
popular disturbances. With the advent of hot, dry August weather
in 1766, optimism about the harvest returned, prices subsided and
social tensions relaxed. It was not until harvest time that the
inferior quality of the grain became apparent. Wheat that had
looked heavy in the ear threshed out coarse and light in weight.

Widely diverging prices reflected the difference in quality between the old and new grains. At this time the price of the
remaining stocks of wheat harvested in 1765 in the shires of
Northampton, Buckingham, Oxford and Warwick reportedly was
52/- and that of new wheat 28/- a quarter. This prompted
Charles Townshend, Chancellor of the Exchequer in the Chatham–Grafton Ministry, to note that the high prices of the old
grain seemed 'to prove the demand abroad and the inadequate-
ness of the stock in hand', and 'the low prices of the new grain
in proportion might be thought to show the defects of the crop
this year in sort although not in quantity'.[112]

Price increases now spread unevenly across the countryside.
Over a period of three weeks or more in September, food prices
rose steeply in all markets of southern England and the Midlands.
The fact that this violent fluctuation was the latest of a series in
1766 accounts in part for the widespread disorders that followed.
Such sudden changes in prices were more disconcerting to the
poor than a steady rise of prices over several months of that year
would have been, for there was some truth in John Pitt's comments to Lord Hardwicke upon the need for stable prices. He
noted that:

The poor knows not how to proportion his labour to his livelihood,
for ninety-nine in a hundred, let times be what they would, would
never get beforehand; while a certainty in his expense would make
a certainty of his labour, and habit would cooperate with necessity.
. . . An equal price of provisions is the best thing for the poor. They
proportion their industry to the accustomed supplies necessary for
a livelihood, whilst a fluctuating price breeds riot and distress.[113]

[112] Charles Townsend to Grafton, 4 September 1766, *Grafton Papers.*
[113] Pitt to Hardwicke, 29 September 1766, Add. MSS., 35607, fol. 310–11.

While such views suggest that the poor received a big enough income for them to set some of it aside for future emergencies, a dubious prospect for agricultural workers by the 1760s, it was probably true that many of the poor could have increased their income marginally in times of gradually increasing prices. In the summer of 1766, there were indications that employment opportunities existed in various parts of the country for those able to move to them. Newspapers, for example, reported harvests in the West Country delayed by a lack of labour.[114] A steady rise of food prices would have enabled workers in a relatively unsophisticated economy to find extra employment or eventually to find cheaper substitutes for more expensive wheaten bread and meat, although the poor were remarkably conservative in their eating habits and reluctant to give up improvements in their diet. Sudden fluctuations in food prices suggested to the poor that there was manipulation of the food supply. It was the assumptions that the poor made about the nature of the food shortage which encouraged their violent responses to the high prices of 'necessaries' in September 1766. These assumptions were derived from long-standing prejudices against one or two interest groups in rural society, which the actions of the government in response to the food crisis confirmed.

In proclaiming the old statutes against middlemen on 10 September, rather than extending the embargo on grain exports on 26 August, the Ministry appeared to indicate that the shortage was created artificially by corn factors and large farmers for their private profit. A general view of an artificial shortage is confirmed by such commentators as James Montagu who wrote to Lord Shelburne of the mutinous disposition of the poor 'who see themselves oppressed with hunger in the midst of plenty'.[115] It is against this background that the mobs' attacks upon middlemen, farmers and anyone moving grain to the urban centres and ports must be viewed. But how remarkable was the government's proclamation of the laws forbidding forestalling, engrossing and regrating? Probably it was not as singular as Professor D. G. Barnes has suggested.[116] In this, they acted entirely predictably. For on at least two earlier occasions of serious food scarcity and riot, the

---

[114] *Public Advertiser*, 18 August 1766 and *passim*.
[115] *Shelburne Papers*, vol. 132, fol. 30.
[116] Barnes, *History of the English Corn Laws*, p. 39.

authorities had done precisely the same thing. The *Gentleman's Magazine* reported that in July 1740 the Lords Justices 'in face of tumults at the dearness of corn' published an order against all 'grain ingrossers' when the prices were above those in the Acts of 5 & 6 Ed. VI.[117] Again in 1756 that journal noted that:

In consequence of several applications to His Majesty by the magistrates of Bristol, Liverpool, Newcastle upon Tyne, and several other seaports relating to the excessive price of corn, the privy council met at the Cockpit and issued a proclamation by which the purchasing of corn for transportation without license is entirely forbidden; the old laws relating to the forstalling and regrating are ordered to be strictly put in execution; and all farmers etc. are enjoined under several penalties to bring their corn to open market and not to sell by sample at their own dwellings on any pretense.[118]

By the 1760s it had become almost a reflex action to remind middlemen of their social obligations. Yet in 1766 the effect of publicly singling out the one interest which had been the object of sustained attack for several years was more serious than ever before. The government's action set off a train of events which resulted in some of the most serious agrarian riots of the century.

By failing on the one hand to take effective action to safeguard the grain supply, and on the other implying that the shortage was artificially created by middlemen circumventing the open market, the Ministry discouraged the movement of grain to market and incited the rioters to attack the scapegoats suggested to them. John Pitt noted the effect of limiting transactions to the open market in a manner more appropriate to the sixteenth rather than the eighteenth century: 'You could not buy an egg or pound of butter but at the tingling of a bell and on a particular spot. This time and place were a direction to the regulators whose violence was a deterance to the country peoples coming in and provisions got dearer.'[119] The *Annual Register*, too, was critical of the government's actions, when in retrospect it observed: 'Many doubted whether this proclamation was well conceived or well timed. It was in some sort, prejudging the question, and declaring

[117] *Gentleman's Magazine*, v (1735) 355.
[118] Ibid., xxvi (1756) 546.
[119] Pitt to Hardwicke, 21 December 1766, Add. MSS., 35607, fol. 341.

the scarcity to be artificial, which experience has since shown to have been but too natural. . . .'[120]

The conclusions of the Ministry were quickly conveyed to the populace. Clerks in country towns publicised the proclamation in local newspapers and constables displayed copies in each hundred, while at the same time magistrates offered to prosecute offenders and encouraged informers at public expense. For example, at Michaelmas the Derbyshire magistrates offered a five-pound reward to informers.[121] Thus the whole rural population was made aware of the government's belief that the shortage was artificial. Soon practical realities forced the abandonment of these regulations against the middlemen for : 'It was apprehended that this measure [the proclamation] would have an effect contrary to the intentions of the council, and by frightening dealers from the markets, would increase the scarcity it was designed to remedy. This was so well felt that little was done towards enforcing the proclamation, and it soon fell to the ground.'[122] But the damage was already done. Not only did the government's action encourage the poor to take matters into their own hands and deal with those responsible for the shortage, it enabled the local authorities for their own purposes to divert attention away from themselves. To understand, however, the character of the responses of all three groups, one must examine the attitudes towards middlemen which developed after the mid-century.

Severe fluctuations in the prices of food, then, precipitated the waves of riots which culminated in the disaffection of the industrious poor of most of southern England and the Midlands in the late summer and autumn of 1766. The major cause of these fluctuations was a natural shortage whose effects were aggravated by government trade policies and the expectations of the people. The Ministry's proclamation of the old anti-middlemen statutes encouraged both the ruling orders and the agrarian poor to blame corn dealers, great farmers and others for the food crisis. This misconception determined the direction of the disorders and their violence.

---

[120] *Annual Register*, ix (1766) 224–6; x (1767) 40.
[121] Derbyshire Record Office, Derby, *Derby Quarter Sessions Order Book* (1766).
[122] *Annual Register*, x (1767) 40.

It is evident that spiralling prices of necessaries were the catalyst which acted upon the deep-seated discontents of more than one interest. Before analysing these, it will be necessary to examine and account for the strong prejudice against middlemen which became apparent after the mid-century and which played a significant part in determining their response to the food crisis of 1766.

2 The Economic and Social Background of the Provincial Hunger Riots

Denunciations of middlemen were common in newspapers, periodicals, pamphlets and private correspondence after the mid-century. The following comments of the Mayor of Guildford were typical of these : 'I hope the above jobbers as is the locusts of the earth will all be silenced and that all the necessaries of life will be obliged to be brought to the public fairs and markets.'[1] Most critics reserved their bitterest attacks for the middlemen of the provisions trade, who were suspected of by-passing or manipulating markets for their own profit at the expense of the consumers. In the 1750s and 1760s there were various indications that public antipathy was reaching serious levels. During the hunger riots of 1756–7, publishers issued a large number of pamphlets which blamed the current shortages and high prices of food on jobbers and provision dealers.[2] Particularly bitter assaults on middlemen followed. Rioters attacked the property of bakers, corn factors and others of the grain trade. They even destroyed Quaker meeting houses attended by Midland corn dealers.[3] After 1764 several Parliamentary committees agreed in blaming middlemen generally for the high prices of all provisions.[4] Pamphleteers once again joined in denigrating middlemen in the 1760s. Newspapers and journals took up the cry too. Occasionally writers rallied to the support of the commercial classes of the privisions trade, but their clamour was muted under the weight of adverse criticism.

---

[1] Thomas Jackman, Mayor of Guildford, to Lord Abercorn, 8 March 1765, *Committee on High Prices of Provisions.*
[2] Barnes, *History of the English Corn Laws*, pp. 32–3.
[3] *Gentleman's Magazine*, xxvi (1756) 409.
[4] Ashton, *Economic Fluctuations in England 1700–1800*, p. 22.

This campaign of vilification reached its peak in the period of scarcity and high prices of 1766–7, when hunger riots were widespread across southern England. Renewed public attacks on the middlemen in the press seriously affected the course of events by influencing the responses of the poor and the rich alike to the food crisis. Thereafter public hostility to middlemen, at least at the official level, declined somewhat. In 1772 Parliament, recognising not only the futility of restrictions on forestalling, engrossing and regrating, but also the serious disruptive effects of their rigid enforcement, repealed the old statutes against middlemen.[5] Popular distrust did not abate as quickly. In 1792 when prices of bread more than doubled, public debate once again focused on middlemen's activities. Edmund Burke and other influential Commons speakers justified middlemen's 'non-productive' activities in a debate on the high cost of food.[6] But this time the authorities made little attempt to prevent the engrossing of supplies, although prosecutions for monopoly were still possible under Common Law. The experience of the 1760s had finally taught that the economy had outgrown such primitive measures of control.

Elsewhere it has been shown that, when faced with rising discontent in early September 1766, the Chatham Ministry took the ill-advised step of proclaiming the laws against engrossing, regrating and forestalling in preference to prohibiting immediately the exportation of grain and permitting the free importation of colonial and foreign supplies. Thereby it broadly hinted to the public that the grain shortage was artificial rather than natural. The newspaper and pamphlet campaign of nearly twenty years had already stimulated the natural loathing of the poor for the miller, the baker and the corn factor. The government's reminder was the final encouragement of the serious riots which followed.[7]

Plainly the timing of the proclamation was unfortunate. The six months' embargo on corn exports ended on 26 August 1766, and in the next four weeks grain, which had been assembled in the ports in anticipation of free exports over several weeks, flooded out of the south and west of England to meet the famine

---

[5] Barnes, p. 41.

[6] Central Library, Sheffield, Fitzwilliam MSS., Burke 18.

[7] Cf. the belief of the French people that a *pacte de famine* had been made to starve the poor (Rudé, *The Crowd in History*, p. 227).

needs of Europe. In this operation corn factors, badgers and other middlemen fulfilled their customary and necessary roles. Pressure to restore the embargo immediately failed in face of the Ministry's doubts about the legality of such action without prior Parliamentary approval and a reluctance to lose traditional markets for grain.[8] The government's action in proclaiming the old statutes against engrossing, forestalling and regrating revealed a lack of appreciation of the functioning of the economy.[9] It was clearly inconsistent to permit large-scale grain exports, while at the same time declaring illegal the very practices necessary to assemble grain shipments not only for abroad but for urban centres such as London.

In part at least, the responses of the authorities and the poor reflected an extraordinary hostility towards middlemen which had become firmly entrenched in the minds of Englishmen about the middle of the century. The student must first seek to understand their prejudices against middlemen if he is to explain the roles of both the authorities and the poor in the serious disorders of the 1760s. Before examining the more specific reasons for the unusually bitter hostility towards this segment of the 'middling sort' however, some consideration of the general social tensions which formed a background favourable to the creation of popular scapegoats for the food crises of the 1760s is desirable.

I

By the mid-eighteenth century the rate of economic growth of England quickened, and in the process intensified stress in society. Rural districts felt the effects of change. The fact that many of these effects were not always to the economic detriment of either the privileged orders or the industrious poor did not lessen their social impact upon traditional communities. Popular antipathy towards middlemen, which was especially evident in the hunger riots of the 1760s, must be viewed against this back-cloth of social flux.

To understand why this period of rapid change occurred after

---

[8] George III to Lieutenant-General Conway, 20 September 1766, *Letters of George III*, ed. Bonamy Dobree (London: Cassell, 1968) pp. 41–2. Harcourt to Jenkinson, 16 September 1766, Add. MSS., 38340.

[9] Barnes, p. 39.

1750 and to appreciate why it had such a profound effect upon important interests, one must go back at least as far as the last decade of the preceding century. In these years rural society was beginning once again to experience important structural change, which resulted from depressed conditions in agriculture.[10] Except in comparatively rare years of poor harvests, animal epidemics, or uncertain trade, the prices of agricultural produce continued to be low throughout the first half of the eighteenth century.[11]

This state of affairs operated for the most part less in the economic interests of the farmers and the landowners than those of the labouring poor.[12] Low returns on agricultural investment tended to drive out the least efficient owners and farmers, and encouraged the close attention to productivity of those who managed to survive. Great landowners, with estates scattered through several regions, were better able to survive than smaller landowners whose interests were more local. Frequently such large landowners were not solely dependent upon their agricultural rents for survival.[13] In times of crisis such men could use capital obtained from industrial or commercial investment to keep their agricultural ventures solvent. By 1700 a trend for less efficient and undercapitalised, small landowners to sell out to larger neighbours was evident.[14]

The engrossing of smaller estates by the owners of great properties had its most immediate effect upon those of the lesser parish gentry and yeomen who were obliged to sell their land. The economic benefit of changing from freehold ownership to leasehold farming was quite evident. Freed from their former obligation to invest in expensive land improvements, in most instances they now farmed their land under long leases, usually for three lives.[15] When the produce market fell to uneconomic levels, as it did particularly in the depression years between 1730 and 1750,

[10] Charles Wilson, *England's Apprenticeship 1603–1763*, Social and Economic History of England, ed. Asa Briggs (London: Longmans, 1965) pp. 141–59.

[11] G. E. Mingay, 'The Agricultural Depression, 1730–50', *Economic History Review*, 2nd ser., VIII (1956) 323–38.

[12] Wilson, p. 249.

[13] H. J. Habakkuk, 'English Landownership, 1680–1740', *Economic History Review*, 1st ser., X, no. 1 (1940) 4.

[14] Ibid., p. 2.

[15] Wilson, p. 252.

they usually found their landlords willing to accept payments in kind or even to forgive the full amount of their rent rather than watch their land deteriorate while they hunted in vain for other tenants.[16] When prices rose sharply in seasons of poor harvests or of animal epidemics, any increased profits went into the pockets of the leaseholders rather than into the hands of the landlords, because rents set earlier in the century when the prices of food were low remained stable.[17] Yet the loss of landownership did represent a decline in social status in a century when property rights were supreme, and hostility to the landed interest was always latent among farmers. Habakkuk has noted that 'hatred of the small squires and gentry for the great lords . . . who were buying them out is the theme of many contemporary plays'.[18]

Not only did the size of estates grow in the first half of the eighteenth century, but so did the size of farms.[19] Landowners preferred to lease their land to large-scale farmers who were most likely to survive the difficult market conditions because they were able to afford the improved methods of husbandry which were slowly spreading through rural England in the late seventeenth century, and to gain the economies of scale. Sir John Fielding claimed that by 1765 rich farmers had swallowed little farms of £70 to £100 per year.[20] Smaller tenant farmers displaced by this trend became labourers working for others or left the land entirely. The growing reluctance of landowners to lease extra land to small freeholders reinforced this tendency towards large farms. Unable to rent land to make their operations economic, these smaller yeomen freeholders sold out and joined the ranks of tenant farmers or labourers.[21]

For his part the rural labourer born to his station in life found the prices of provisions low and life relatively easy for most of the first fifty years of the eighteenth century. Throughout southern England the woollen cloth trade prospered and the disruptive

---

[16] Chambers and Mingay, *The Agricultural Revolution 1750–1880*, pp. 20–1.

[17] Ibid., p. 47.

[18] Habakkuk, p. 12.

[19] *Gentleman's Magazine*, xxxv (1765) 85. See also Sir John Fielding, 'Observations of Prices of Provisions', 5 February 1765, *Committee on High Prices of Provisions*. W. G. Hoskins, *The Midland Peasant*, cited in Wilson, p. 250.

[20] Fielding, op. cit. See also *Gentleman's Magazine*, xxxv (1765) 85.

[21] Wilson, p. 251. Habakkuk, p. 15.

effects of the expansion of Yorkshire's worsted cloth industry still lay in the future.[22] Agricultural workers could thus supplement their earnings under the domestic system. Population growth was not yet putting pressure upon village communities. Migration to the metropolitan area, to other growing urban centres and to the colonies, at least for the single man, provided a safety valve through which discontents could be dissipated.[23] While those who descended from owner-occupiers to tenant farmers or labourers faced a difficult period of adjustment, only in rare years of crisis was there severe strain in rural society in the first fifty years of the eighteenth century.[24] In a few years of poor harvests and high prices, food riots did break out, but they were scattered and short-lived compared to the chronic hunger riots of the second half of the century.[25]

During the 1750s there was a marked increase in rural tensions, which in the first instance was precipitated by a change in the terms of domestic trade in favour of agriculture. Essentially the causes of this phenomenon were threefold: population growth after 1740[26] and movement to the developing industrial regions and urban centres,[27] government victualling contracts during the Seven Years' War,[28] and natural disasters such as bad harvests and animal epidemics.[29] The most evident and immediate result of this swing in favour of agriculture was higher prices for produce, which continued throughout the remaining years of the century and affected the well-being of all rural interests.[30]

[22] *Victoria County History, Gloucestershire* (1907) II 160: The greatest prosperity in the cloth trade of the west was between 1690 and 1760. Mantoux places the serious competition of the northern towns after 1790 (*The Industrial Revolution in the Eighteenth Century*, rev. ed. [London: Methuen, 1966] p. 264).

[23] Phyllis Deane and William A. Cole, *British Economic Growth, 1688–1959, Trends and Structure*, 2nd ed. (Cambridge: Cambridge University Press, 1967) p. 115. Campbell, 'English Emigration on the Eve of the American Revolution', pp. 1–20.

[24] Habakkuk attributes the absence of comparable social tension to that preceding the Civil War to the transference of land to Conservative elements ('English Landownership', p. 5).

[25] Robert Featherstone Wearmouth, *Methodism and the Common People of the Eighteenth Century* (London: Epworth Press, 1945) *passim*.

[26] Deane and Cole, pp. 93–4.

[27] Ibid., pp. 111–21.

[28] Ashton, *Economic Fluctuations*, p. 60.

[29] Ibid., p. 22.

[30] Ibid., pp. 181–2.

Inevitably the benefits of greater agricultural profits were spread unevenly through rural society. Those landowners able to pass on to their tenants the increasing burden of taxation and other costs were able to share in the growing profitability of commercial farming.[31] Others who were committed to long leases at low rents were not able to gain from the favourable movement in agriculture. They had to bear the cost of higher taxation to pay for the war, as well as the growing weight of welfare necessitated by the rising cost of living.[32] Where great landowners frequently benefited indirectly from higher prices resulting from natural disasters, smaller owner-occupiers were more vulnerable to bad harvests or epidemics of cattle and sheep, which were more frequent in the 1750s and 1760s than hitherto. Many of these lesser landowners were forced to sell their land.

The trend towards larger agricultural units, evident earlier in the century, accelerated after 1750. Although the upswing in the produce market enabled some small landowners who might otherwise have been forced off their land to survive, many found it desirable to sell out to their more efficient and better capitalised neighbours.[33] The increasingly profitable markets encouraged investment in estate improvement and better farming techniques. Scientific farming methods known and practised on a limited scale in the previous century such as the marling of sandy soils, surface drainage, varying crop rotations, selective breeding of animals and the rest, spread rapidly after the mid-century owing to the efforts of great landowners and progressive farmers. As a result, productivity and ultimately rents rose, especially on the light soils.[34] Once again the greatest advantages went to the large-scale operators. Wealthy landowners were best able to take advantage of the lower interest rates on capital necessary for the development of estates.[35] At the same time engrossing landowners were able to exact a more nearly economic rent upon land recently acquired because they were not bound by lengthy leases.

[31] Victoria County History, Wiltshire, IV (1959) 62.

[32] G. E. Mingay, English Landed Society in the Eighteenth Century (London: Routledge and Kegan Paul, 1963) pp. 83–4.

[33] Habakkuk, 'English Landownership', pp. 1–17.

[34] E. L. Jones, 'Agriculture and Economic Growth in England, 1660–1750: Agricultural Change', Journal of Economic History, XXV (1965) 11.

[35] Ashton, An Economic History of England: The Eighteenth Century (London: Methuen, 1955) pp. 40–1.

Such landlords now preferred tenancies at will rather than lease-holds for several lives.[36]

Generally owners of land close to expanding urban centres or industrial districts were in the best position to benefit from the growth of commercial farming by the 1760s. Owners of corn lands, however, benefited from the ease of transportation of grain, and even those distant from the coasts or urban centres shared in the new prosperity. Less fortunate were landowners tied to long leases or those whose land produced less readily transported commodities. These only participated in the benefits of the com-mercial farming boom after long delays during which leases ran out or transportation systems developed. Those who managed to survive this extended lean period were obliged to watch resent-fully the success of other landowners and the growing affluence of great tenant farmers. One anonymous polemicist expressed the resentment of many of the landed interest when he wrote, 'Was it ever thought of, in the original institution of agriculture, thât the husbandman, who rented £300 per annum should be enabled to live better than his landlord, who had no other income?'[37]

Although by the 1760s in counties such as Wiltshire, one of the most disaffected counties in the hunger riots of 1766, the landed interest was beginning to pass on to their tenants the weight of increased taxation in the form of higher rents, the situation of the larger farmer in southern England after the mid-century was favourable.[38] As was the case among landowners able to exact an economic rent, the greatest beneficiary of expanding commercial farming was the large-scale farmer located close to London or some other growing urban centre, or to developing industrial regions such as the Midlands, Lancashire, the West Riding or Hallamshire. Even where he was located at a distance from markets, the great corn farmer was prosperous by the 1760s. His product was relatively easily transported on the improved rivers and canals.[39] When exports were permitted, he gained from ex-

---

[36] Chambers and Mingay, p. 47.

[37] Anonymous, *A Letter to the House of Commons in which is Set Forth the Nature of Certain Abuses Relative to the Articles of Provisions* (London: J. Almon, 1765) p. 34.

[38] There was an acceleration in the growth of agricultural output after 1750 (Deane and Cole, p. 75). See also Wilson, p. 254.

[39] Mantoux, *The Industrial Revolution in the Eighteenth Century*, p. 125.

port bounties; when scarcity at home shut off foreign markets, he benefited from the enhanced prices in home markets.

The large agricultural producer in the 1750s and 1760s gained from other factors besides expanding markets for farm produce. Among these were the natural disasters which occurred with greater frequency after 1750 than in the early years of the century. Generally the weather had been favourable to raising crops and animals for most of the first fifty years of the eighteenth century, but now adverse weather conditions resulted in several years of poor harvests. Particularly noteworthy were the seasons of 1751, 1756 and 1766 which produced light harvests and high prices.[40] In other years shortages of fodder crops adversely affected cattle and sheep. In 1762–3 there were very small crops of hay in Wiltshire which led subsequently to much slaughtering of calves and cattle shortages and high prices in following years.[41] Unusual weather in the early 1760s led to a fluctuation in the supply of acorns which was later reflected in a shortage of hogs and high meat prices.[42] The high barley prices of 1763 resulted in the breeding of fewer pigs and the subsequent high prices of pork commented on in the House of Lords' report on high meat prices in 1765. Animal epidemics increased in the middle years of the century and raised the prices of meat.[43] Cattle murrain was widespread between 1745 and the mid-1760s and decimated many herds.[44] Natural disasters such as these were often sufficient to destroy small farmers whose interests were purely local. But a great farmer, some of whose stock and crops survived, profited from the scarcity and enhanced prices of meat and grain. He too was better able to meet the rising costs of poor relief, which was the concomitant of widespread distress caused by high food prices, than his smaller competitor.[45]

The conspicuous consumption in which many of the prosperous

---

[40] Ashton, *Economic Fluctuations*, pp. 20–2.

[41] Extract of a letter from Mr Frawd, Gentleman Farmer, of Brixton Deveril in Wiltshire, to the Lord Bishop of St David's, 2 February 1765, *Committee on High Prices of Provisions*.

[42] Ashton, *Economic Fluctuations*, p. 22.

[43] Butler to Lord Leigh, 20 February 1765, *Committee on High Prices of Provisions*: reported sheep rot in Warwickshire, Leicestershire and Northamptonshire.

[44] Ashton, *Economic Fluctuations*, p. 20 and *passim*.

[45] Ibid., p. 42.

great farmers engaged after 1750 attracted much comment. Frequent complaints in newspapers and journals of the period testify to the resentment of the wealthier farmers' social ambitions, which was shared by those above and below them on the social scale, and to the mounting social tension. Many of these farmers were now adopting pastimes hitherto the prerogatives of gentlemen: hunting, driving in carriages, employing maidservants for their newly-leisured wives and sending their children to public schools and universities. Newspapers testified satirically to their changing tastes by noting advertisements in country journals such as the following : 'Wanted by a gentleman farmer, a complete ploughman, who can also drive a pair of horses on occasion. N.B. He must know how to dress hair after the London fashion, and if he knows farriery so much the better.' Such social pretensions encouraged many in the belief that landowners should force their tenant farmers to occupy themselves fully with their traditional labours and should discourage them from expectations inappropriate to their stations in life. Typically one writer urged a return to less sophisticated living in these terms : 'Let our farmers be farmers, that is, let them live by labour, let their sons follow the plough, and their dames and daughters attend the dairy, and not change a country life for the foibles of a court, and to become imitators of nobility.'[46] Other correspondents advised landowners to raise their rents sufficiently high to ensure the desirable close attention of their tenants to the care of the land and to discourage upward mobility. Often writers attacked the practice of amalgamating smaller farms into larger farms. One such critic expressed his concerns in the following sense :

. . . every landlord ought to keep in view the supporting the rank of the industrious farmers and not endeavour to raise them into the higher station of yeomanry nor to say gentry—this will be done no way so effectively as by assortments of farms of proper size for public good and fixing such rents as will keep up a tenant's attention and industry which are the best security a landlord can hope for.[47]

But the direction of social mobility in agriculture was more commonly horizontal or downward by the 1750s than upward, and many successful tenant farmers resented their inability to acquire the social prestige and political influence that landowner-

[46] *Westminster Journal and Political Miscellany*, 21 May 1768.
[47] *Shelburne Papers*, vol. 132, fol. 89–90.

ship carried. Despite their growing wealth, large farmers found it increasingly difficult to buy land. During the eighteenth century the land market tightened. Even before the mid-century, despite the unprofitability of agricultural land, the price of estates rose steadily during and after the 1720s.[48] The legal device of strict settlement, introduced in the previous century, was becoming popular amongst the leading members of the landed interest as a means of preventing the alienation of their property by future, less provident generations.[49] The net effect of this growing practice of entailing estates was a tendency to freeze landownership, and thereby increase the price of land and make upward social mobility more difficult for tenant farmers and others. At the same time that this reduction in the fluidity of existing estates was occurring, there was growing competition for available land. Men enriched by commercial ventures in India, America and the West Indies, together with a growing number of successful industrialists, sought the social status and political influence that landownership conferred.[50] Wealthy men could still buy land throughout the century, but by the 1760s it was becoming increasingly expensive and difficult. Frustrated in their desire to rise into a higher social class and to enjoy status and power commensurate with their wealth, many tenant farmers resented what they regarded as the unproductive role of the aristocracy and gentry. In times of crisis these latent resentments came to the surface. Only on very rare occasions were the farmers able to make common cause with the rural poor against the landed interest. When this form of 'class' polarisation took place, it threatened the supremacy of the aristocracy and particularly the gentry in the localities. One such occasion, which will be dealt with below because it had such significant influence upon events in 1766, occurred in 1756–7 when the local authorities attempted to implement a new, unpopular Militia Act in a period of unrest caused by high food prices.[51] More commonly the larger farmers, rather than the landowners, felt the resentments of the poor after the mid-century.

[48] Ashton, *Economic Fluctuations*, p. 94.

[49] Mingay, p. 32 *et seq.*

[50] Wilson, cites H. J. Habakkuk, 'The Land Market in the Eighteenth Century', in *Britain and the Netherlands*, ed. J. S. Bromley and E. H. Kossmann.

[51] Western, *The English Militia in the Eighteenth Century*, pp. 290–302. Rockingham to Newcastle, September 1757, Rockingham MSS., RI-105.

The rising cost of necessities aggravated the grievances of the lower orders against the great farmers. Popular antipathy against this interest approached that directed towards the middlemen after the mid-century. Farm wages failed to keep pace with the cost of living especially after the Seven Years' War. Although in many of the southern counties wage rates did rise slightly in the 1760s, they actually went down in the west of England.[52] At the same time the prices of wheat and meat rose steeply. In most of the southern half of England the poor ate wheaten bread and therefore their standard of living was seriously affected by the increased cost of wheat.[53]

The tendency towards consolidation of estates under fewer owners and the creation of larger farms had an indirect effect upon the life of rural labourers too. In some cases it merely resulted in a change of landlord, whose contact with the agricultural worker was minimal. In many instances, however, new ownership meant scientific management, with its emphasis upon maximum productivity. This resulted in a greater specialisation of labour on the farm. Instead of working at the whole range of farming tasks, a labourer now was expected to concentrate upon being an expert cowman, for example. Frequently a new breed of tenant farmers dedicated to farming efficiency discouraged their workers from part-time occupations which competed for their time and energy. Many farmers now actively discouraged their labourers from engaging in woollen cloth production under the 'putting-out' system, which had traditionally provided the poorly-paid farm worker with a useful supplement to his income.[54]

Even where tenant farmers did not prevent their labourers' participation in the domestic system, the rural poor by the 1760s were finding less opportunity to supplement their wages by working in the cloth trade. The competition of the woollen worsted industry of the West Riding was forcing fundamental reorganisation upon the old woollen cloth regions. Increasingly the cloth centres of the West Country and East Anglia were switching to the production of fine woven cloths in the late

---

[52] G. D. H. Cole and Raymond Postgate, *The Common People, 1746–1946*, 4th ed. (London: Methuen, 1966) p. 76. See also Gilboy, *Wages in Eighteenth Century England*, p. 134.

[53] Charles Smith, *Three Tracts on the Corn Trade and the Corn Laws* (1766) p. 182.

[54] Ashton, *An Economic History of England*, p. 115.

1760s.[55] Such products were less suited to cottage production by relatively unskilled farm workers with little machinery at their disposal. As a result, farm workers experienced a decline in their standard of living by the 1760s and were more dependent upon agricultural wages.

Much of their resentment they directed towards their employers. In times of crisis especially, the rural poor attacked the property or persons of farmers who stressed profitability at the expense of traditional rights. Thus in the Norwich riots of October 1766 a rural mob attacked one yeoman farmer for 'had not the old rogue whipped the gleaners from his fields'.[56] In the hunger riots of 1766 the farms were the objects of searches by mobs, and farm produce heading for markets or the ports was intercepted.[57]

But it was another aspect of the reorganisation of landowner-ship which after the mid-century seriously affected the conditions and status of the rural poor. The growing profitability of com-mercial farming, the availability of 'cheap' money and the shrink-ing of the land market encouraged the enclosure of both cultivated and waste lands at an increased pace after 1760. The precise effect of the enclosure movement upon the agricultural labourer is impossible to measure, and certainly its accelerated growth was only beginning to have an impact by the late 1760s. But certain tendencies were already evident.

Much debate has in the past centred around whether or not enclosures caused the depopulation of rural England and pro-vided the labour force to operate the new factories. The consensus of historians now is that where enclosures for pastoral purposes took place, they did drive men off the land. Thus in the Mid-lands, where sheep runs and cattle pastures increased in the 1760s as a result of enclosures, many of the displaced poor crowded into the weaving villages of Leicestershire to set up as stocking weavers, an occupation which required little capital. The insani-tary conditions created by this sudden influx of population, the undernourishment which resulted from the high food prices after the Seven Years' War and led to serious epidemics and over-crowding—all these created tensions which manifested themselves

[55] Julia de L. Mann, 'Textile Industries since 1550', in *Victoria County History, Wiltshire*, IV (1959).

[56] Norwich Record Office, *Depositions*, 1766.

[57] Wearmouth, *Methodism, passim*.

in extensive food riots throughout that county in 1766.[58] But where enclosures developed for the purpose of creating arable farm lands, they increased rather than decreased job opportunities. Nor was the actual division of land by Parliamentary commissioners under the various acts of enclosure performed with as little concern for the traditional rights of the lower orders as was once claimed by Fabian socialists who wrote on the agricultural revolution in the nineteenth and early twentieth centuries. The commissioners often gave compensation for rights which could not be substantiated by documentary proof.[59] The increased productivity which resulted from the rationalisation of agriculture was also beneficial for it enabled Britain to feed her growing population better than would have been the case had small-scale farming continued to predominate into the nineteenth century. Yet the economic and social changes which occurred inevitably created tensions in rural society. Individuals suffered as a result of the redistribution of formerly communal land. Cottagers and squatters particularly lost important supplementary sources of income when common land was enclosed. Any compensation they may have received for such losses was inadequate. Robbed of a cushion against outright destitution, which the ability to raise one or two animals and cultivate a small kitchen garden gave, these members of the lower orders often became wage-earning labourers solely dependent upon their farmer-employers for their subsistence and much more vulnerable to fluctuations in the price of food.[60] Paradoxically the years of food crisis and rioting, to which the reorganisation of land ownership contributed, stimulated further the tendency towards large estates, and an increase in the number of enclosure bills followed each of the years of serious scarcity and high prices of the second half of the eighteenth century.[61]

Although the resentment of the poor against enclosures in the 1760s had not yet built up to the level it was to reach later in the century, some riots against enclosures did occur, notably in

[58] W. G. Hoskins, 'The Population of an English Village 1086–1801—A Study of Wigston Magna', *Provincial England* (New York: Macmillan, 1963).

[59] W. H. Chaloner, 'Recent Work on Enclosure, the Open Fields and Related Topics', *Agricultural History Review* (1954).

[60] Chambers and Mingay, pp. 97–8.

[61] G. E. Mingay, *Enclosure and the Small Farmer in the Age of the Industrial Revolution*, Studies in Economic History (London: Macmillan, 1968) p. 20.

Northamptonshire.[62] But most of the resentments against social and economic change were expressed more indirectly in the agrarian hunger riots of this decade.

As a result, then, of the reorganisation of landownership and the spread of scientific farming practices, by the 1760s a clear threefold division of agrarian society into landowners, tenant farmers and labourers was beginning to emerge. The roles of each of these three interests were gradually becoming more clearly defined than ever before. Many tenant farmers had ceased to own any freehold; many landowners ceased to farm the land they occupied; and many labourers lost traditional rights to communally-owned property. While this neat division into rentier, manager and rural proletariat did not occur overnight, and certainly was not complete by the 1760s, the trend was already evident. The tendency towards a polarisation of rural society in times of crisis caused the ruling orders concern.

Yet these three interests were nowhere near as homogeneous as might first appear. There was as much difference within interests as between them. This was partly because social orders and economic interests did not coincide. The landowners ranged in status from the lesser parish gentry to the great county gentry and aristocracy.[63] The farming interest still included owner-occupiers such as yeomen freeholders, as well as tenant-farmers who varied in size from mere subsistence farmers to wealthy leaseholders who were scarcely distinguishable in standards of living from fairly well-to-do gentry. The labouring interest included those dependent solely upon agricultural wages, together with farm labourers who still cultivated their own small pockets of land, raised animals upon the common, and earned supplementary income from family involvement in the domestic system of cloth production.

Of all the divisions within rural groups after the mid-century, that within the landed interest was to have the greatest impact upon the riots in 1766. By the second half of the century there was apparent a decline in the common outlook among landowners which had been characteristic of earlier times. Now the lesser parish gentry and the great county families shared little beyond a common rentier relationship with the land. Local antipathy towards the growing metropolitan influences was much greater at

[62] *Gentleman's Magazine*, xxxv (1765) 441.
[63] Habakkuk, 'English Landownership', p. 3.

the parish level than at the county level. Parish gentry were particularly resentful of the disruptive effects on local markets of the activities of London food buyers.[64] On the other hand, county gentry were more cosmopolitan in outlook. Their economic interests and social connections were more diversified than those of their poorer fellows. Frequently as representatives of the county or some borough, the great gentry spent months in London and were closer to the national government than parish gentry. More important, a common educational experience for lesser gentry, county gentry and the aristocracy was disappearing in the early years of the century. Great families were now more frequently sending their sons abroad on the Grand Tour rather than putting them to study the classics at Oxford or Cambridge. The Inns of Court in London, too, were ceasing to attract the sons of the more exclusive families in the landed interest.[65] This social gulf which appeared during the early years of the century when great landowners were engrossing land at the expense of the lesser landowners was a serious cause of estrangement between segments of the landed interest. The poorer gentry often found the cost of living high, rents too low, and taxes crippling after the mid-century. Such problems affected them more than they did the great landowners who had other resources besides rents, and who could buy out owner-occupiers and raise rents under new leases. The lesser gentry who provided the government at the local level felt themselves increasingly estranged from county families who often represented as Lords-Lieutenant the national government's policies. The initial failure of the two levels of government to co-operate in suppressing the rioters created a very serious situation during the hunger riots of 1766.

Much of the rivalry within interests resulted from regional disparity. In the dairy region of Wiltshire, for example, farmers and landowners enjoyed lower returns than the landowners and farmers of adjacent corn lands. Dairy farmers in this county were close to subsistence and had a reputation for turbulence which stretched back into the Civil War period.[66] Hardly distinguishable from labourers in standard of living, these men doubtless joined

[64] Ibid., p. 3.
[65] Lawrence Stone, 'The Ninnyversity?', *New York Review of Books*, 28 January 1971, pp. 21–9.
[66] *Victoria County History, Wiltshire*, IV (1959) 64.

in the protests against high food prices in the 1750s and 1760s. Although farmers in Gloucestershire, another county seriously disaffected in the 1766 hunger rioting, were supplying the developing cheese and butter markets of London by the mid-century, most perishable goods could not readily arrive in the metropolis from the West Country and other interior regions lacking access to rivers or the sea. Small farmers who lacked ready access to population centres frequently enjoyed only a marginal existence. The poor conditions of many farmers in the West Country and in parts of Norfolk, for example, were evident in the practice of paying rents and wages in kind rather than in money in the 1760s.[67]

This uneven growth pattern of agriculture led to an increasing gulf between the more successful and less affluent members of the same interest groups. Rivalry within interest groups was as stress-producing as competition between interests. The stimulation applied by the growth of produce markets after the mid-century served to exacerbate economic disparities. The social tensions produced by such economic conditions were nowhere more apparent than in the popular attitude towards middlemen during the hunger riots of the 1760s.

II

Dislike of middlemen was not peculiar to the second half of the eighteenth century. The frequency with which terms like 'engrossing' are used pejoratively in journals or pamphlets during and after Tudor times, and the sizeable body of legislation to curb 'abuses' in the markets by Tudor and Stuart Parliaments testify to the universal distaste for attempts to corner the supply of any commodity whether it was food, land or merely the tools essential to a particular trade.[68]

The quality and price of food were always of immediate concern to the poor. Traditionally they were quick to express resentment at short-measure or adulteration of provisions. Because

[67] *The History of the City of Norwich: From Earliest Records to the Present Time* (Norwich: W. Allen, 1869) pp. 346–7.

[68] Coal undertakers had gained a monopoly of shovels and used their control over the essential tools of the coalheavers' trade to dominate the labourers ('The Present State of the Coalheavers', dated 1768, William L. Clement Library, Ann Arbor, Michigan, *Sydney Papers*).

bread was the staple diet of the poor in southern England, bakers were the object of suspicion.[69] Fluctuations in the economy and changes in its structure from time to time intensified these dormant suspicions, and the accumulating legislation of Tudor and Stuart times reflected popular resentments. The growth of London, particularly from the sixteenth century on, created a need for 'engrossing' food supplies. The number and activities of middlemen increased.[70] Local hostility towards salesmen, whose competition for food raised prices in local markets, was inevitable. To this sharpened natural antipathy was added parochial suspicion of alien metropolitan influences, which the centralising tendencies of the Stuart and Cromwellian governments exacerbated.

An examination of the philosophy, provisions and application of the statutes against the abuses of middlemen is an essential prerequisite for an understanding of the role of this commercial interest in the later eighteenth century.

Sir John Fielding, a leading metropolitan magistrate, ably summed up the philosophy behind the various paternalistic acts which sought to protect the rights of consumers when he noted in 1765 : 'As to the articles of luxury in life, they may be left open to exorbitant gain in the seller without much injury to society. But as to the absolute necessaries of life, as they relate to the useful part of mankind, the legislature should constantly interpose to prevent extortions and monopolies.'[71] A justice and a gentleman, Fielding reflected the attitudes of his brother magistrates and their class when he spoke in favour of the old 'moral economy'. Not only did gentry feel a sense of *noblesse oblige* towards their poor, but they were always sensitive to circumstances which threatened the peace of the countryside and the towns. They frequently suspected middlemen of profiteering in order 'to get an estate'.[72] High prices of food and suspicions of exploitation created dangerous resentments among the lower orders. Their experience of Common Law taught the leaders of rural society 'that provisions of all kinds—alive or dead—ought

---

[69] Fay, *The Corn Laws and Social England*, p. 4.

[70] Norman Scott Brien Gras, *The Evolution of the English Corn Market from the Twelfth to the Eighteenth Century* (Cambridge, Mass. : Harvard University Press, 1915) pp. 208–9.

[71] Fielding, 'Observations of Prices of Provisions', 5 February 1765.

[72] William Payne to Lord Abercorn, 12 February 1765, *Committee on High Prices of Provisions*.

to be sold in the open market'.[73] The administration of bounty payments on grain exports, the bread assize and the fixing of wages were all predicated on prices of provisions finding their own levels in 'free' markets. A correspondent in the *Gentleman's Magazine* confirmed the wider implications of food prices when he wrote of the government's duty to regulate the profits on the necessaries of life, for these prices were '. . . the natural regulators of the prices of labour of all kinds'.[74] As one historian has asserted, the concept of a 'just price' went back at least as far as medieval times.[75]

While there were already significant voices raised in favour of *laissez-faire* by the 1760s, the paternalism of the old 'moral economy' would have found general acceptance in the nation at large. It was the feasibility of implementing such principles in the increasingly sophisticated economy that raised doubts.

Protection of the consumer was enshrined in law. In addition to the general protection from monopoly given under Common Law, several statutes sought to control specific abuses. Probably the most frequently cited were 5 & 6 Edward VI, cap. 14, and subsequent acts which amended their provisions. These statutes defined the illegal practices of forestalling by purchasing 'any merchandise, victual, etc. coming towards any market or fair, or coming towards any city, etc.', or making 'any motion for the inhancing of the prices . . .'; regrating by buying and selling again within four miles of a particular fair or market; or engrossing by 'buying, contracting or promise taking, other than by demise, grant or lease of land or tithe' any corn still growing in the fields or any other grain with the intention of selling again.[76] While these offences were rather narrowly defined, later terms such as forestalling, engrossing and regrating had a wider application, and 'stood almost as a single phrase for unpopular manipulation in time and place of the people's food'.[77] Other acts specifically restricted middlemen in the livestock and meat trades. By a Tudor statute no one was to have more than 2000 sheep at one

[73] Thomas Brock, Clerk of the Peace and Town Clerk of Chester, to Lord Abercorn, 11 March 1765, ibid.
[74] *Gentleman's Magazine*, xxxiv (1764) 27–8.
[75] Rose, 'Eighteenth Century Price Riots'.
[76] Fay, p. 53.
[77] Ibid., p. 54.

time.[78] Under another act of Charles II, no butcher was to offer for sale 'live oxen, steers, runts, kine, calves, sheep or lambs'.[79]

After the Restoration, the middleman's legal position improved greatly: 15 Car. II, cap. 7, 'made it lawful for all and every person, when corn did not exceed a specified price, to buy in the open market, and to lay up and keep in granaries or houses and sell again such corn' provided they were not 'forestalling nor selling in the same market within three months after buying thereof'.[80] As a result of this relaxation, the true wholesaler-corn merchant was said to have grown up in the metropolitan area with the blessing of the government, who wished to see the Dutch corn dealers replaced.[81] By the mid-century his equivalent in the meat trade, the carcass butcher of Smithfield, had emerged to dominate the livestock trade, despite a well-organised lobby of retail butchers from Newgate, Clare and other London markets, and sporadic prosecutions.[82]

The application of the laws against middlemen generally ebbed and flowed with the strength of popular resentment of middlemen in the food trade which varied with economic conditions and the prices of provisions.[83] But there were a number of reasons, besides the generally low prices of food and relative prosperity of the rural poor, why justices in the first half of the century turned a blind eye to contraventions of the paternal statutes and market by-laws.

Legal recourse from the manipulation of market supplies was difficult. The requirement that individuals give information before magistrates discouraged the prosecution of offenders under the Tudor and Stuart statutes. The informer was always the object of universal loathing. The *Gentleman's Magazine* noted that the 'odious name of informer' discouraged the enforcement of laws

[78] 25 Henry VIII, cap. 13.
[79] 15 Car. II, cap. 8.
[80] Fay, p. 54.
[81] Ibid.
[82] Ray Bert Westerfield, *Middlemen in English Business, Particularly between 1660 and 1760* (New Haven, Conn.: Yale University Press, 1915) p. 217. *Committee on High Prices of Provisions* (March 1765).
[83] Two years of high prices when prosecutions for forestalling and regrating were undertaken were 1757 and 1765 (see *Gentleman's Magazine*, XXVII [1757] 479; XXXV [1765] 95).

against lightweight bread.[84] Occasionally the legal records recount the brutal retribution exacted upon informers months after the suppression of a riot.[85] More commonly such acts are hidden under the charge of common assault. Frequently the mobs diverted personal violence towards their own members, rather than against the authorities. To avoid such revenge, as well as to share the costs of litigation, gentry and others formed private associations to prosecute those suspected of market manipulation in times of crisis. In December 1766 Pitt reported to Hardwicke that in the West Country gentlemen were forming groups to prosecute forestallers.[86] In a similar fashion, committees of textile manufacturers prosecuted workers for embezzlement, defective spinning, wetting or oiling cloth and delay in returning materials. Associations to prosecute offenders under the old paternal statutes among the commercial and farming interests lapsed in times of plenty, which were the rule rather than the exception in the first half of the century.

Although it was difficult to discourage forestalling, engrossing and regrating, there is little doubt that the authorities could have significantly reduced such activities had they wished to do so, simply by making the seller as liable in law as the buyer.[87] This reluctance to pursue wholeheartedly the suppression of market abuses hints at the dilemma of the ruling orders. Both the aristocracy and the gentry resented the growing affluence of many middlemen and great farmers, but were aware that the apparently beneficial results of the corn bounties were only obtainable with their help. A grain ship's cargo had first to be engrossed by a jobber or corn factor. There was a contradiction in passing the Corn Law of 1689 in the interests of the producer and enforcing a variety of statutes in the interests of the consumer. In practice the enforcement of the latter was nominal.

[84] *Gentleman's Magazine*, xxvi (1756) 557. See also M. W. Beresford, 'The Common Informer, the Penal Statutes and Economic Regulations', *Economic History Review*, 2nd ser., x (1957–8) 221.

[85] *Gentleman's Magazine*, xli (1771) 189–90, notes the brutal slaying of a witness in a weaver's trial several months after the trial.

[86] John Pitt to Hardwicke, 21 December 1766, Add. MSS., 35607, fol. 341. *Gloucester Journal* (September 1757) noted meetings of gentry to form associations to prosecute forestallers.

[87] William Payne to Lord Abercorn, 12 February 1765, *Committee on High Prices of Provisions*.

Even when individuals were willing to swear out infor-
mations against offenders under the various anti-middlemen
statutes, prosecutions were not easy. Many illegal market activities were difficult to prove. Price fixing in the back rooms of inns
might be strongly suspected, but charges were difficult to substantiate. Farmers and corn dealers could transact much business
discreetly over a private dinner. Critics claimed that great farmers
sold by 'these latent contracts' grain at a minimum of 3d per
measure below the open market or 'peddling price', which was
established by bakers buying from lesser farmers, who brought all
their grain to market. The poor, unable to buy in gross, had their
bread assized on the basis of the peddling price 'to the double
profit of the baker and mealman'.[88]

The temptation to circumvent markets in this way was great.
Plainly, it was a great inconvenience for the larger farmers to
bring great quantities of grain, meat or other bulky commodities
to the local markets. With the development of commercial farming by the mid-century to feed the growing urban population and
satisfy overseas markets, there was increasing encouragement to
sell by sample. In 1765 this means of forestalling the market was
the most common cause of complaint to the Lords' Committee
inquiring into the high prices of food. One witness before the
House of Lords' Committee inquiring into the high prices of food
in 1765 complained that farmers 'bring a bushel or two or three
in the public market, and keep ten at their inn, and on the
appearance of a scanty market, if they can raise it to their price
will produce by degrees or else they'll produce a sample and by
that bring to the Baker's house from 20 to 50 or 100 bushels at
an agreed price, so that the poor can't buy any'.[89] The decline of
the small farmer producing for local markets which accelerated
after 1760 with the rapid extension of enclosures and the amalga-
mation of small farms by 'monopolising farmers' aggravated the
problem of dwindling sales of provisions in the open market.

Prevention of selling by sample required the concerted action
of the officials of all markets. Where adjacent markets permitted
this type of forestalling, the clerks of other markets were eventually forced to accept it also, despite the statutes and local by-laws,

---

[88] R. Wright, Town Clerk of Warwick, to the Right Honourable, the Earl
of Abercorn, 9 March 1765, ibid.
[89] Thomas Miller to Lord Abercorn, 7 March 1765, ibid.

or face ever-declining revenue from the reducing use of their facilities.[90]

Perhaps the worst consequence of such sales by sample was that they helped to obscure the actual market price of food. As one local official wrote, '. . . the publick cannot judge whether there is plenty or not, nor do they know what price three fourths of it [the corn] is sold for'.[91] The common practice of dealers shipping commodities like barley and oats directly to the maltster's storehouse, or sending them to badgers for shipment by water to great markets like Bristol, compounded this problem. Another witness before the House of Lords' Committee on High Prices, referring to the Exeter market, claimed that 'the prices of barley and oats cannot be ascertained with the same precision [as wheat] as they do not fall under the public cognizance'.[92] Grain prices quoted at this time in the *Gentleman's Magazine* and elsewhere were misleading because often only bulk buyers could buy these grains at such prices. Uncertainty about the actual prices of provisions bred suspicion of middlemen and fears of famine in times of scarcity. The poor, never far from subsistence levels, were more susceptible than ever to market rumours. In periods of crisis they were increasingly prone to take revenge upon the middlemen, for, they asked, was it not better to hang than starve?

Another factor that added to the confusion and made the enforcement of the old statutes difficult was the variety of measurements which made it impossible to compare prices even in neighbouring markets, and therefore to judge if middlemen were taking unfair profits.[93] The problems arising from this confusion

[90] A correspondent in December 1765, referring to the justices' failure to lower prices by prosecuting butchers, higlers, graziers, dealers in cattle and other engrossers, cited the case of vigorous enforcement in one city causing supplies to 'desert the market' (*Gentleman's Magazine*, xxxv [1765] 613–6). Wright noted that magistrates were obliged to overlook infractions of the by-laws in Warwick (R. Wright, Town Clerk of Warwick, to the Right Honourable, the Earl of Abercorn, 9 March 1765).

[91] Thomas Jackman to Lord Abercorn, 8 March 1765, *Committee on High Prices of Provisions*.

[92] Jacob Rowe to Lord Abercorn, 9 March 1765, ibid.

[93] 'One measure throughout the kingdom would likewise be of great service, for the Welles, Shepton Mallet, and Somerton [markets] sell by large measures or six packs to the bushel, yet there is a difference in all three . . .' (Thomas Miller to Lord Abercorn, 7 March 1765). Wright complained of the deceptions of farmers and millers using various measures (R. Wright to Lord Abercorn, 9 March 1765). Corn sold at Guildford market 'by almost as many measures as farmers . . .' (Thomas Jackman to Lord Abercorn, 8 March 1765).

of measures are apparent in the following entry in the *Order Book* of the Gloucestershire magistrates :

Whereas, by reason of the neglect in putting the several acts of Parliament for ascertaining the measures of corn in execution, great inconvenience and losses have happened and do happen to the King's subjects in general and to the poor in particular, in as much that, through the uncertainty of measures, the sellers of corn, and grain do not well know what measures to bring to market, nor can the magistrates settle the assize of bread according to the price of corn as it is sold in the market whereof the poor have not so much bread for their money as they ought to have. . . .[94]

By the 1760s there was an effort to standardise measures, and most markets were adopting the Winchester grain standard of eight bushels to the quarter. But it is evident from the statistics the *Gentleman's Magazine* in 1766–7 tried in vain to collect regularly from its volunteer correspondents across the country, that practices continued to vary, and a quarter might still contain eight, eight-and-a-half, nine, or ten bushels of grain. The House of Lords' Committee in 1765 was careful to ask clerks of markets to indicate the size of measures in local use.[95]

Nor was the size of the measure the only variable. People complained that the shape was critical too.[96] The wider the measure, the more the tendency of grain to settle, and thus a shallow, wide container was said to hold more grain than a tall, narrow one of the same cubic capacity. These discrepancies increased when larger, globular commodities like potatoes were to be measured. Even when markets used standard-shaped measures, weight was a more significant factor than size for such commodities as grain. Several writers argued that there was less danger of fraud when weight measures were used, for weight was a better indicator of quality.[97] Corn produced in wet seasons was frequently larger in the ear, but lighter and coarser than grain produced in dryer

---

[94] Gloucestershire Record Office, Gloucester, *Quarter Sessions Order Book*, no. 9 (1766–80) 15 July 1767, D214/B10/4. See also William Beveridge, *Prices and Wages in England from the Twelfth to the Nineteenth Century*, 2nd ed. (London: Frank Cass & Co. Ltd, 1965) I *passim*. As early as 1709 the Gloucestershire justices drafted a petition to Parliament for a standard measure of corn (*Quarter Sessions Order Book* [1766–80] D214/B10/4).

[95] *Committee on High Prices of Provisions* (March 1765).

[96] J. Tomlinson to Lord Scarsdale, 9 February 1765, ibid.

[97] *Considerations on the Exportation of Corn*, p. 64.

seasons. For other commodities such as cheese, beef or veal the price per pound quoted in the press gave little indication of quality. The confusing number of variables in quantity and quality made comparisons next to impossible. Their social superiors told the poor constantly in the 1750s and 1760s that the middlemen were cheating them, but the authorities seemed powerless to prevent the evil. Not surprisingly in times of scarcity, the dispossessed hit out blindly at their 'oppressors'.

Probably the most significant reasons for the growing toleration of middlemen in the early eighteenth century related to the rapidly-changing economic reality. Increasing urbanisation on the one hand, and industrial specialisation on the other, demanded a more sophisticated system of distribution of food than ever before. In 1757 the *Gentleman's Magazine* observed that the laws against forestalling, engrossing and regrating were 'so antiquated and the circumstances and manner of living of all ranks of the people so alter'd, that a vigorous execution of them would rather contribute to famish than feed in many places great numbers of the poorer sort'.[98] By the mid-century the growth of outports such as Liverpool and industrial centres such as Leeds, Birmingham and Manchester complemented the expansion of the older population centres of London, Norwich and Bristol.[99] Manufacturing regions like the West Riding of Yorkshire and parts of Lancashire were rapidly losing their self-sufficiency in food production.[100] They were becoming increasingly dependent on more distant agricultural regions. The activities of middlemen were essential to the satisfaction of the needs of both the cities and the new industrial regions of the north and the Midlands. Significantly, it was the metropolitan magistrates who were among the first to recognise the value of middlemen in the increasingly complex marketing system. The preamble of the act of 1772, which finally repealed the statutes against forestalling, engrossing and regrating, underlined this newly-realised dependency when it noted the great distress visited 'on the inhabitants of many parts of the kingdom and in particular the cities of London and Westminster by their enforcement'.[101]

---

[98] *Gentleman's Magazine*, xxvii (1757) 129.
[99] Deane and Cole, pp. 111–22.
[100] Westerfield, p. 130.
[101] Fay, p. 55.

But it was not just economic considerations which militated against the application of protectionist regulations in the interests of the consumer by the early eighteenth century. Political realities, too, were antipathetic. 'Small government' concepts and practices of the century replaced the Stuart and Cromwellian stress on centralisation. After 1689 authority and influence returned to the localities.[102] Not until the nineteenth century, when the leaven of Benthamism mixed with the general reaction to the problems of a new industrial society, was there a strong movement towards centralisation again. Paradoxically the leaders of the localities, the gentry and the aristocratic landowners, in times of crisis, tried to return to the old restrictions of the former 'moral economy'. They failed to recognise that the political decentralisation that they favoured rendered impotent their piecemeal efforts to control the abuses of middlemen.

Finally, good harvests provided enough food to feed the population and to export a surplus to Europe, which helps to explain the more permissive attitude adopted by the authorities towards the activities of middlemen in the first half of the eighteenth century. Plentiful and cheap food favoured the poor, if not the landowner and farmer. Only in the years 1709, 1727–8 and 1740 were hunger riots serious,[103] and therefore was there any pressure on the government to reimpose the old statutes against the middlemen. By the mid-century, this interest was well entrenched, and it had long since become essential to the increasingly sophisticated economy. What then caused the role of middlemen to be re-examined in the 1750s and 1760s?

Several factors account for the renewed questioning of the value of middlemen after the mid-century. These factors include the controversy over the Corn Laws, which erupted in a large number of pamphlets published between 1751 and 1756; the emergence of new classes of affluent middlemen due to the accelerated growth of the metropolis and the exigencies of war; rivalry between interest groups; public concern at the cost of provisions between 1756 and 1758 and between 1764 and 1769; and the stresses of a period of rapid economic change.

---

[102] John M. Norris, *Shelburne and Reform* (London: Macmillan, 1963) p. 292.
[103] Rudé, p. 36. Ashton, *Economic Fluctuations*, pp. 17, 144, 147. Barnes, p. 32.

The Corn Laws first attracted widespread attention among pamphleteers in the early 1750s. Before examining some of the arguments centred upon these laws, it is necessary to comment briefly on their major provisions.

The first corn bounty appeared in Charles II's reign. Under 25 Car. II, cap. 1, a reward in the form of a bounty on grain exports was paid to landowners for the subsidies granted by Parliament to fight the first Dutch war.[104] Parliament in 1689, however, passed the Corn Law which operated for most of the next century.[105] Partly enacted to encourage the development of European grain markets and to persuade the landed gentry to accept the Revolutionary Settlement, the Corn Law of 1689 formed the basis on which bounty payments were made down to 1773.[106] The historian of the Corn Laws, D. G. Barnes, has noted that this act represented an important change of emphasis from consumer to producer needs in economic policies.[107] Under its provisions, exportation was not prohibited at any price; at a price of 48/- a quarter the poundage duty was to be 1/- and under 48/- a bounty of 5/- was to be paid on every quarter exported; duties on wheat imports were to be at the rate of 1/4 when the price was more than 80/-, 9/- when the price was 53/4 to 80/-, 17/- when it was 44/- to 53/4, and 22/- below 44/-.[108] Similar provisions and proportionate rates were provided for other grains, except oats, which was not controlled.[109] The regulations discouraging the importation of grains were necessary concomitants of the bounty provisions. It was important to prevent re-exports from gaining bounty payments intended to stimulate home production and encourage the landed interest. During the period under consideration, before 1772 that is, attention centred on the bounty aspect of the Corn Laws.[110] Thereafter the protectionist trade aspects of the laws held the attention of critics and supporters alike. In the pamphlets published in the 1750s and 1760s,

---

[104] Barnes, p. 10, citing the *Quarterly Journal of Economics*, xxiv (1909–10) 419–22. See also *Considerations on the Exportation of Corn*.
[105] Barnes, p. 11.
[106] Ibid.
[107] Ibid.
[108] Fay, p. 29.
[109] S. and B. Webb, 'The Assize of Bread', pp. 196–218.
[110] Barnes, Chapter III.

controversy centred around whether greater agricultural production resulted and whether the cost of living of the poor increased because of the Corn Laws. The relative merits of the debate are not directly relevant here. Probably the long-term result of the bounty programme was to increase the amount of grain produced, but the clumsy machinery and the shortcomings of a marketing system in transition ensured that in sudden crises the grain stocks were not preserved and the poor suffered grave hardships.[111]

The existence of statutes tells only part of the story and one must turn to Privy Council records and elsewhere to assess the executive's use of them. Embargoes on grain exports and the free importation of colonial or foreign grains became more frequent after the mid-century. There were also various measures of self-help taken by merchants and gentry. The formation of associations to provide the poor with subsidised grain helped reduce tension in periods of high prices.[112] Such measures were a tacit admission of the failure of the self-regulating machinery of the Corn Laws. Government interference with the free-working of these laws occurred at times of actual or anticipated shortages of food, and were of short duration before the 1760s. Previously the government had occasionally suspended exports in times of critical grain shortage, for example in the years 1709, 1740 and 1741. Between 1756 and 1773 suspension was frequent.[113]

The Corn Laws first came under hostile scrutiny as a result of high government expenditure rather than a scarcity of grain, which was the ostensible cause of the agrarian disorders of 1756–7. Heavy grain harvests in 1749–51 resulted in a very large exportation of grain in those years, which embarrassed the government with the need to find large sums of money to pay the export bounty.[114] Because of a legal decision declaring the South Sea Company's dividends the first claim on Tunnage and Poundage, the original statutory source for bounty payments, the Ministry

---

[111] Ibid., p. 30.

[112] Henry Roper to the Earl of Abercorn, 8 March 1765, *Committee on High Prices of Provisions*; *Gentleman's Magazine*, XXVIII (1758) 42; and *Shelburne Papers*, vol. 132, fol. 59 and 63.

[113] *Three Tracts on the Corn Trade and the Corn Laws.* See also Barnes, p. 23.

[114] Barnes, pp. 23–4.

had to honour the corn debentures out of treasury funds. Indirectly the burden fell on the landowners, who already paid a land tax inflated by wartime demands.[115] It was in the early 1750s that many gentry began to take an interest in economic reform.

Between 1751 and the 1756–7 hunger riots, newspapers and pamphlets debated the value of the Corn Laws.[116] The supporters of the bounty system claimed that it encouraged expanded grain production which, while enriching the country in normal times from foreign exchange, provided a food reserve that could always be diverted to home consumption in periods of crisis. They further argued that the system kept down the price of grain sold at home through the economies of large-scale production. Not only were the rich landowners said to benefit, but the poor also. Because all trade was interdependent, everyone, pamphleteers said, benefited from the consequent prosperity of agricultural growth : merchants, manufacturers, tradesmen and seamen, as well as the agricultural interests.[117] Writers drew evidence in support of their contentions from the Eton College records of purchases in the Windsor market over a number of years in the early eighteenth century.

The direct beneficiaries of the bounty system were the grain farmers, their landlords, the corn factors in the export trade and shipowners. Less obvious vested interests were the supporters of sugar distillery. Sugar planters, West Indian factors, brandy merchants, sugar bakers, brokers and brewers all wished grain diverted towards the export trade.[118] Dutch and French interests, too, benefited from exporting gin and brandy to England, and wished to divert English grain away from distillery and into exports. All these groups exerted a powerful influence in support of the Corn Laws.

---

[115] Mingay, *English Landed Society*, p. 80 *et seq.*

[116] Barnes, p. 24 *et seq.*

[117] There was also some debate in periodicals regarding distillery from corn. The issue of the corn export bounty became mixed with debate on distillery (*Gentleman's Magazine*, xxvii [1757] 71; xxix [1759] 630; xxx [1760] 18, 23–4).

[118] A correspondent, 'J.M.', referred to vested interests among the supporters of sugar refinery: sugar planters, West Indian factors, brandy merchants, sugar bakers, brokers and brewers (*Gentleman's Magazine*, xxx [1760] 18).

Opponents of the bounty noted the limitations of the statistical evidence advanced by the supporters of the Corn Laws. They were wholesale figures and averages taken at only two dates in each year. These critics urged that the bounty was a means of subsidising foreign competitors by providing their employees with cheap grain at the expense of the British taxpayer, while at the same time English industry suffered from the enhanced price of labour due to dear necessaries, and poor rates increased to enable the indigent to pay the higher price of bread.[119] Some critics directed their attacks on the operation of the bounty system itself, which they claimed was wide open to abuse. The opponents of the bounty system were for the most part from the manufacturing interest, but Irish and American agricultural interests found the laws discriminatory too.

Both the landed interest and the corn middlemen, as the apparent beneficiaries of the corn policies, were the culprits in the eyes of their critics, as is apparent from the following typical comment: 'Thus evidently appears the blessed advantage of exportation. The public are taxed on the common necessaries of life, at the rate of nine million annually; only to support the landed interest and jobbers in grain.'[120] Yet it was the landed interest, as the dominant influence in the legislature, which carried the responsibility for measures which at their inception had been intended to benefit them, and whose interests were publicly associated with them. While many of the rising industrial interest regarded the Corn Laws as a type of outdoor relief for the gentry and aristocracy, the extent to which the landed interest profited is questionable. Adam Smith believed that the Corn Laws were established from a mistaken sense of the real interests of the country gentry, and that the chief beneficiaries were the corn merchants.[121] Certainly the landed interests were scarcely less divided in their attitudes to the Corn Laws in 1766 than they were in 1846, however united they might appear in public.[122]

---

[119] *Public Advertiser*, 4 February 1768.
[120] *St. James's Chronicle*, 26–28 May 1768.
[121] Fay, p. 15.
[122] Ward concludes that in 1846 '. . . traditional loyalties, sentiment, personal and family attachments, protectionist anger and Whiggish rationalism exercised more influence than did economic considerations . . .' (J. T. Ward, 'West Riding Landowners and the Corn Laws', *English Historical Review*, LXXXI [April 1966] 271–2).

While correspondents in the press generally regarded the landed interest as a monolithic force, individual landowners complained that they were an 'unconnected tribe' that might be treated anyhow and 'made to endure what such contemptible herds as vintners or tobacco merchants would not hear of'.[123]

Many landowners and farmers did not benefit from the Corn Laws directly; indeed the smaller ones who lacked diversification often suffered higher costs because of the bounty system. Outside the heavy corn-growing regions, many landowners rented out cattle and sheep country.[124] The costs of livestock farming increased with the higher cost of cattle feed, which the bounty system caused. Where bread increased in price because of exports stimulated artificially by the bounty on grain, poor rates too rose. Many landowners shared in the unpopularity of their interest as a result of the Corn Laws without profiting from their operation. As noted earlier in the 1760s lesser landowners were often experiencing economic and social difficulties which aggravated further the impact of the bounty system.

Response to the Corn Laws was not determined solely by considerations of economic interest, real or imagined.[125] As local magistrates, landowners were willing to support the suspension of grain exports in times of emergency. They were as much concerned with maintaining public order as with personal economic advantage.[126] The fact that such suspensions only occurred after prolonged exports had created a severe shortage of grain in the country and after the bounty debentures were issued probably reflected more on the Ministries who were reluctant to lose traditional markets in Europe or stir up powerful lobbies by 'premature' action.[127] Parochial prejudice against metropolitan influence also played a role in determining local attitudes to national legislation and to the drawing power of the 'Great Wen'.

---

[123] James Harris to Hardwicke, 3 October 1766, Add. MSS., 35607, fol. 316.

[124] Westerfield, pp. 444–5, lists the following counties where corn was not a significant product by 1762: Oxford, Buckinghamshire, Surrey, Middlesex, Devon, Warwickshire and Lincolnshire.

[125] Ward, 'West Riding Landowners and the Corn Laws'.

[126] Rose, p. 292.

[127] A writer raised a legitimate question when he asked, '. . . Should we lose our market for corn abroad what other commodity have we to bring a balance in our favour?' (*Gentleman's Magazine*, xxxv [1765] 195).

Publicly landowners paid lip service to the popular myth of the unity of the landed interest. Not having read Adam Smith on the subject, most Englishmen equated the Corn Laws with the interests of the landed. But landowners who doubted the efficacy of the bounty system could not bring themselves to attack it openly. They preferred to express their opposition more obliquely by criticising the jobbers in grain and other middlemen, without whose activities grain could not be exported in large quantities and the benefits of the bounty gained. Professor Barnes has found the large number of pamphlets published on the causes of scarcity and the high prices of food after the 1756–7 food riots extraordinary. No such flood had followed the earlier agrarian protests in 1709, 1727–8 or 1740, he noted. Nor did he see any connection between these pamphlet attacks on middlemen and the earlier controversy over the corn bounty in the years between 1751 and 1756, because in that case he believed the corn would have been attacked.[128] Yet it would seem reasonable to suppose that both the *laissez-faire* opponents of the Corn Laws, who were drawn largely from the ranks of the manufacturing classes interested in lowering their costs of production, ending British subsidisation of foreign competitors and freeing foreign markets from reciprocal tariffs, and those landowners who were disadvantaged by the bounty system, would welcome a shift of attack to the middlemen. The landed interest was too strong for a frontal assault in the eyes of industrialists, while the powerful social loyalties of landowners prevented their appearing to attack publicly their own interest. The poor, too, could readily identify the ubiquitous middleman as the culprit. Even those of the landed interest who benefited directly from the Corn Laws saw the chance of diverting public hostility away from themselves. As will be seen, this anxiety to find a scapegoat was reinforced by the sense of isolation and vulnerability that both the aristocracy and the gentry felt as a result of the class feeling generated against them with the riots over the new Militia Act and high food prices in 1756–7.[129] The dangers of such tactics do not seem to have occurred to them until the extent of the disaffection of the lower orders became apparent in the autumn of 1766.

[128] Barnes, p. 32.
[129] Western, *The English Militia in the Eighteenth Century*, p. 299.

The Corn Laws, then, encouraged the growth of middlemen, and the controversy centring upon the operation of the export bounty system stimulated a general antipathy towards middlemen and large farmers which had lain dormant in the early years of the century. For their own purposes, various interest groups found it convenient to attack publicly the role of the middlemen in the 1750s and 1760s. After the mid-century there were many targets for anti-middlemen prejudice.

By this time many corn factors, jobbers, salesmen, carcass butchers and other middlemen were conspicuous because of their affluent standard of living.[130] The rapid growth of London particularly, and the opportunities of large wartime contracts, had enriched many. With the growing sophistication of the economy, changes occurred within the provision trade. Some middlemen declined in importance, while others rose to positions of influence and even of monopoly. Carcass butchers and salesmen from Smithfield market appear to have established a monopoly over the London meat trade at the expense of retail butchers and drovers by the mid-century.[131] At least part of the popular distrust of established middlemen was stimulated by dispossessed lesser middlemen. Millers, who reportedly changed into wholesale mealmen or flour merchants, persuaded farmers of the advantage of immediate cash payments in place of more speculative processes of shipping direct to distant markets. By giving small advances to lesser farmers, these middlemen were said by their critics to dictate prices.[132] Because of large victualling contracts with the navy to supply pork, distillers became large-scale breeders of pigs, which fed on the fomented mash. In 1756, for example, the *Gentleman's Magazine* reported a contract for 10,000 hogs at 1000 per week signed by distiller contractors and the victualling office. Such contracts discouraged embargoes on distilling in times

[130] Westerfield, p. 217 *et seq.*, examines in detail the complexity of London middlemen.

[131] The terminology of the marketing system was often confusing. 'Retail' and 'cutting' butchers were synonymous terms. Other names described different functions at different times. There was, too, much overlapping of occupations. Carcass butchers overlapped with salesmen and graziers, for example. *Gentleman's Magazine*, xxv (1755) 294, complained of the graziers' control of the meat market. Fielding, too, noted the dominance of retail butchers by carcass butchers ('Observations of Prices of Provisions', 5 February 1765).

[132] *Gentleman's Magazine*, xxviii (1758) 424.

of scarcity, especially when wartime food demands were pressing.[133] Other contractors undertook to supply oats and other grains much in demand for the English and allied forces in the Seven Years' War. The conspicuous consumption of these newly-enriched middlemen excited the resentment of other social orders. Writing to Lord Abercorn, Chairman of the House of Lords' Committee on High Prices, in 1765, William Payne noted that one carcass butcher had publicly declared his opposition to lower prices and 'his intention to get an estate'.[134] Some of the poor, who had lived on diets of inferior meat and bread in the armed forces, no doubt remembered old scores when they returned to civilian life after 1763, and found prices high. They readily attributed the food scarcity to the manipulations of the middlemen.

Opportunities for speculation in food certainly existed after the mid-century, and doubtless contributed to higher prices. Because of the ease of transportation and storage, grain was well suited to speculation. Lower interest rates in Holland at various times before the mid-1760s enabled speculators to ship grain to the Low Countries, where it could be stored more cheaply than in England, until the prices at home rose sufficiently to justify its return.[135] The export bounty from the home port paid the costs of transportation and the dealers gained large profits from selling when the market was most suitable.

The stage of development reached by the English marketing system in the 1750s also favoured speculation. The economy was in an interim stage. It was rapidly moving from a system of purely local markets towards a more distinctly national marketing system. Communications and transportation were improving rapidly with the improvement of river navigation and the beginning of canal building. The economy still consisted of a number of port-economies, the largest and most dominant of which was London. Arthur Young discovered it was not possible to relate the prices of food directly to the distance from the capital, for each port exerted its own influence over its hinterland. 'Cirencester market governs those [prices] of Tetbury, Hampton, Stroud, Northleach, Fairford, Letchlade, etc., therefore the variation is

---

[133] Ibid., xxvi (1756) 625.
[134] William Payne to Lord Abercorn, 12 February 1766.
[135] Fay, p. 15.

not worth computation', observed the Bishop of St David's.[136] Nevertheless where markets had ready access to water, they were responsive to price changes in London and elsewhere. One correspondent argued the sensitivity of South Wales to price changes in these terms: '. . . The same may be said for all other grains [besides oats] even in more distant places as long as there is a demand and a market, and the seas and rivers are navigable.'[137] Price disequilibrium in one port economy was ultimately reflected in the others.[138] It was the time lag in this process that enabled speculation to succeed.

Even the most distant markets felt London's influence. Reports in the *Gentleman's Magazine* frequently mention the effect on local markets of heavy buying by London salesmen. Cattle sold as far away as northern Scotland and Durham eventually found their way to the London market at Smithfield.[139] Carcass butchers were said not to wait for the market but go down as far as Northampton to buy up cattle and sheep and 'by reason of their large stocks keep up the prices of meat'.[140] Sir John Fielding calculated that within five miles of the city milk was used in its natural state, five to fifty miles milk was used in suckling calves for veal. He noted that, while some butter was made within these distances, there was little cheese. Cheese and butter were produced at more distant centres according to the nature of the land.[141] London had extended its tentacles across the countryside in search of provisions at least from Tudor times, but the rapid growth of the metropolis by the mid-eighteenth century had increased the dealers' scale of operations.

Despite the relationship, then, that existed between different port economies and markets, the rate of response to disequilibria elsewhere was such that local market manipulation was possible. Corn bounty regulations combined with eighteenth-century administrative incompetence had always made possible the defeat

[136] Bishop of St David's to Lord Bathurst, February 1765, *Committee on High Prices of Provisions*.

[137] *Gentleman's Magazine*, xxviii (1758) 278.

[138] C. W. J. Granger and C. M. Elliott, 'A Fresh Look at Wheat Prices and Markets in the Eighteenth Century', *Economic History Review*, 2nd ser., xx, no. 2 (August 1967), 257–65.

[139] Westerfield, p. 187.

[140] Thomas Addison, Butcher, Clare Market, in evidence before House of Lords' Committee (*Committee on High Prices of Provisions* [March 1765]).

[141] Fielding, 'Observations of Prices of Provisions', 5 February 1765.

of the spirit of the regulations, if not the letter. The growing sophistication of the economy and the interim stage of marketing made speculation more widespread and serious. By the mid-century jobbers and corn factors, with the aid of improved communications, were able to develop accurate intelligence systems to report market conditions throughout England and Europe too.[142] At a time when legislation assumed the free market play of prices for the setting of wages and the prices of bread, as well as for fixing price thresholds to regulate the flow of grain out of the country, dealers were said to defeat the self-regulating mechanism by agreeing among themselves to buy up grain on a particular day in a certain market. The effect of such heavy buying was to push up the price of grain above the level at which bounty was payable on exports through the nearby ports (above, that is, 48/- per quarter). Thereupon the corn factors waited until the next market day, when by mutual consent they abstained from buying. Thereby they forced down the price below 48/-, which enabled them to export their accumulated stocks and receive the bounty. This interference in the free play of the markets, critics said, prevented the intended operation of not only the Corn Laws but the assize of bread, the opening of the ports for importation of provisions, and 'in general every power relative to the prices as well as the weight, measure and quality of human subsistence'.[143] Because of this type of manipulation and the slow response of other markets, especially those inland, food prices were often prohibitively high in the interior industrial regions at a time when coastal prices were low enough to enable grain to flood out of the country, a factor which greatly contributed to the resentment of the poor and which was relevant to the hunger riots of 1766.[144]

That the government had to suspend the Corn Laws and place embargoes on all grain exports in times of economic crisis is a clear indication that the 'self-regulating' price mechanisms did not work. In 1773 the Pownall Act changed the price thresholds, but by this time the country had become a net importer of grain and the protectionist aspects of the laws became most relevant to the high cost of necessaries.[145]

[142] See Westerfield, *passim*.
[143] Thomas Brock, Clerk of the Peace and Town Clerk of Chester, to Lord Abercorn, 11 March 1765, *Committee on High Prices of Provisions*.
[144] *Considerations on the Exportation of Corn* (1766).
[145] Barnes, p. 43.

There was, therefore, some basis in truth for accusing some middlemen of affecting the food supply by manipulating markets for the purpose of speculation. Exact measurement of the effect of their actions on food prices is impossible. Probably contemporaries exaggerated it, but certainly it is an important reason for the general unpopularity of middlemen. It probably accounts for the recommendations for the reimposition of controls on middlemen's activities which came from the various Parliamentary committees of inquiry into high food prices of the 1760s.

If the rapidly-increasing numbers of middlemen and the expansion of their activities was stimulating the general antipathy towards middlemen which had remained dormant during the early years of the century, the events of 1756–7 intensified rivalries between various interests. In these years a natural shortage had caused the prices of grain to rise steeply, and disorders took place in widely scattered regions of England.[146] The chief disaffected areas were in the north and the Midlands, but the west and south also witnessed mobs of several hundred protesters.[147]

The response of the Privy Council to these outbreaks and to the several petitions of large seaports such as Bristol, Liverpool and Newcastle-upon-Tyne about the excessive prices of corn was predictable. Meeting at the Cockpit, they issued a proclamation prohibiting the purchase of corn for transportation without a licence, ordering the strict enforcement of the old statutes against forestalling, engrossing and regrating, and requiring all corn to be sold in the open market and the end of sales by sample.[148] Clearly the opportunities for profiteering were greater in times of shortage than in periods of plenty, and the government wished to discourage the withholding of supplies. Yet the implication was plain: the shortage was artificial and the middlemen were exploiting the consumers by manipulating the food supplies.[149]

---

[146] The *Gentleman's Magazine*, xxxvi (1766) 557, noted the distress of the poor of the Vale of Evesham, 'the finest vale of corn in the world', because grain was drained off via navigable river to Bristol.

[147] *Gentleman's Magazine*, xxvi (1756) and xxvii (1757) *passim*.

[148] Ibid., xxvi (1756) 546.

[149] *Two Letters on the Flour Trade, and the Dearness of Corn: By a Person in Business* (London, November 1766) pp. 18–19, complained that public attacks on forestalling, engrossing on dealers and mealmen incited a 'spirit of mobbing' by spreading fears of famine and hindering the regular supply of the markets.

Although the results of the government's action were not as serious as they were to be in 1766, by publicly blaming the middlemen and then failing to regulate them effectively the government incited the rioters to take matters into their own hands. At first they emulated their Tudor ancestors and forced the sale of provisions at what they thought were 'just prices'. Later, as they became more exasperated, they attacked the property of farmers and middlemen suspected of hoarding food. Occasionally quantities of grain were destroyed and frequently the transportation of supplies to the urban centres was impeded. Thus, the government's action was ultimately self-defeating for it discouraged the movement of provisions and created even greater shortages than had at first existed.

As noted earlier, the hunger riots of the 1750s were more serious because of their association with widespread protests against a new Militia Act.[150] These particular riots are dealt with at length elsewhere. But for the purpose of this analysis, it is important to note that the middle-class farmer and the rural poor were pitted against the upper-class gentry and aristocratic landed interest.[151] Certainly, the coincidence of the militia riots with the protests against high prices encouraged the landed interest to vent its resentment for the rough handling it received at the hands of the rural populace against the middle-class farmers and provision dealers.

Much of this resentment was reflected in the pamphlets and newspaper articles which poured forth when calm returned in 1758. Interpretations of the earlier events were many, but the most common scapegoat was the middleman, who, writers claimed, had created an artificial shortage. Corn buyers, kidders, laders, broggers, carriers, flourmen, bakers, brewers, distillers and taverners were all singled out as offenders. Many urged the traditional solution of enforcing the old Tudor and Stuart statutes. The reprinting in 1758 of the *Book of Orders* which expounded government policy towards middlemen between 1586 and 1630 was hardly coincidental. In the event it failed to achieve its purpose of persuading the government to return to a policy of rigorous regulation.[152] Neither did the authorities implement a

[150] Western, p. 300.
[151] Ibid.
[152] Gras, *The Evolution of the English Corn Market from the Twelfth to the Eighteenth Century*, p. 207.

scheme for establishing public granaries. One student of the Corn Laws has seen this as a reluctance of the government and the landed interest to revive the old bureaucratic system of the crown.[153] Probably it would have gone against the political spirit of the times to return to the centralism of the Stuart and Cromwellian periods.

A period of favourable harvests and lower prices followed the crisis years of 1756–7. The conditions of the poor improved for several years, until in 1763 poor harvests and cattle epidemics forced up prices again. The findings of several Parliamentary committees which examined the causes of high food prices in the next three years are important for they confirmed the popular opinion about the role of middlemen in raising the cost of living for the poor. Because many of the committees' recommendations concerned particularly conditions in the metropolis, they not only coloured the views of the national government on the causes of popular unrest but they also revealed economic changes that were happening. These changes are of interest because they formed a background to both the agrarian disorders and the pre-industrial riots of the later 1760s.

Popular antipathy to middlemen in London was rather less than in the provinces. The most vociferous critics of the larger middlemen were the lesser middlemen whom they were displacing as the metropolitan marketing system grew more complex and demanded greater infusions of capital. This development was more evident in the meat trade than the grain trade by the 1760s.

To judge from the reports made to the House of Lord's Committee of 1765, the London carcass butchers had come to dominate the metropolitan meat markets in the previous decade. The well-organised retail butchers' lobby which made detailed submissions to the Committee claimed that a great increase in the numbers and influence of these large-scale whoesalers had occurred in the previous three years. The accelerated growth of London in the 1750s and 1760s due to the war, a shortage of livestock due to animal epidemics, and a post-war recession probably account for this sudden increase in the numbers and influence of the carcass butchers.

An examination of the operation of these wholesale meat dealers shows the complexity of the problem of controlling

[153] Barnes, p. 33.

middlemen and suggests why governments found it best to ignore tacitly their monopolising tendencies, while publicly paying lip service to the principles of the old 'moral economy'.

The operation of the carcass butchers centred primarily about Smithfield, the market for live cattle and sheep. Here they dealt in animals bought from distant farms or local markets. By keeping animals on the marshlands near London or dispersed through adjacent counties, critics said, they avoided prosecution and played their stock into the market when prices were most favourable. Thus by the use of their superior purchasing power and by manipulation, the carcass butchers came to dominate the smaller cutting butchers in Clare, Newgate and other markets of the metropolis.[154] These retail butchers complained that they were no longer able to buy small drifts of cattle for slaughter for their own customers but were forced to buy on credit from their new masters, who obliged them to accept bad meat along with the good at the price of good meat.[155]

The disgruntled cutting butchers blamed the carcass butchers for the high prices of meat in the 1760s. Sheep and cattle, they claimed, were bought in Smithfield, driven out to nearby fields to pasture or feed on turnips, and later sold when prices were high. Farmers, they said, had ceased to come to London with their stock, and dealt with Smithfield salesmen directly. Occasionally a farmer dissatisfied with the salesmen's prices would try to market his own animals, whereupon the carcass butchers reportedly withheld from buying to force down the prices in Smithfield market. Such independent farmers found it a costly waste of time to by-pass the salesmen, who travelled from farm to farm. Thereafter the farmer sold to Smithfield outriders 'who thus secured the power to starve the public'.[156] The cutting butchers charged that their rivals could afford to sell meat cheaper but would not. They asserted that the recent drop of $\frac{1}{2}d$ per pound of meat was due to the interest of the House of Lords' Committee of 1765 in meat prices.[157]

Certainly the operations of some butchers were large-scale and their investment heavy. Under the strict application of the old

[154] *Committee on High Prices of Provisions* (March 1765).
[155] Ibid.
[156] *Gentleman's Magazine*, xxxiv (1764) 334.
[157] *Committee on High Prices of Provisions* (March 1765).

Tudor and Stuart statutes they were illegal. Some reportedly kept as many as 1000 sheep at one time, and sold as many as 100 per day. They travelled as far as Northampton to buy stock. Typical of this interest group was Benjamin Cherry of Hertford, who sold 10,000 sheep at Smithfield in one year. One witness bitterly asserted that 'he buys them of the farmers, is a judge of the intrinsic value, and plays them into the market to the injury of the public and the very ruin of ourselves and families'.[158] His indictment before Justice Fielding in 1765 had little apparent effect on Cherry's conduct. Edmund Burke, nearly thirty years later, writing of the extent of Cherry's activities, observed that he and others 'had kept a line of circumvallation twenty miles around London where were 40,000 sheep always ready to beat down the market'.[159] According to evidence given before Sir John Fielding, Cherry had at that time 4240 sheep in flocks averaging 250 on various farms throughout Hertfordshire.[160]

The attacks of the cutting butchers did not go unanswered. Salesmen and others maintained that carcass butchers were a great convenience to the public. They provided a means to sell the numbers of cattle and the large quantities of meat needed by the street hawkers and small retail butchers on credit.[161] They noted that the cutting butchers had no facilities for slaughtering in the city, which was forbidden by law. They denied they were a monopoly. Provisions, they claimed further, taken at a medium over the previous twenty years were very little dearer than the twenty years before them; they estimated not more than $\frac{1}{2}d$ per pound. They asserted that cattle and sheep were turned out to grass because of a lack of buyers, for they would have preferred to sell at a lower price than keep the stock. The recent higher prices and shortages of meat were due to greater demand and to rot in recent wet seasons. They cited the reduced number of animals sent from Lincolnshire fens as proof of this.

[158] R. Studley to the Duke of Bedford, Lord President, 29 January 1765, ibid.
[159] Fitzwilliam MSS., Burke 19.
[160] Cherry's servant confessed his master sold 10,000 sheep in Smithfield in one year (*Committee on High Prices of Provisions* [March 1765]). See also 'Return of Cherry's Sheep in Hertfordshire from Jan 1 1765 to 18 Jan 1765', ibid.
[161] Evidence of salesmen (ibid., March 1765).

John Bryant, a salesman of pork and lamb, offered a particularly vigorous defence of his role in the marketing of meat.[162] He operated in both Smithfield and Newgate markets, and employed farmers in Somerset and others of the western counties. He denied there was forestalling on the roads. He claimed that carcass butchers were not new, and were valuable in supplying hawkers with meat. High prices were not due to carcass butchers; rather they were the result of wet seasons in 1760 and 1763. Bryant claimed that the smaller breeds of pork affected prices also. He noted that cutting butchers could not handle the quantities of meat that carcass butchers bought. Without pasturelands near London, cattle and sheep would be ruined by lengthy journeys, he observed. Cattle and sheep were returned to pasture to keep an equality of price, for farmers required more certainty in value and they would have lost if they had to sell as soon as the cattle reached market. Finally, Bryant believed that salesmen, by informing farmers of price variations and seasonal fluctuations in demand, controlled the flow of cattle and sheep to London.

Whatever the merits of the arguments on each side of the debate, public questioning of the role of middlemen indicated that there were important changes happening in the marketing system by the 1760s which were setting up social stresses. These changes accelerated after the mid-century and disrupted the hierarchical structure of the commercial system. Criticisms of emerging 'monopolies' came from within and without the retail and wholesale trades in the 1750s and 1760s. The 'abuses' in the meat trade primarily related to the London markets, but certainly similar objects of criticism could have been found in all the port economies of the eighteenth century. Popular criticism of middlemen was greatest in the countryside which supplied the urban populations. London's freedom from hunger riots in 1766, despite high food prices, suggests the importance of popular hostility towards middlemen in the counties where hunger riots were widespread in this year.

Resentment of middlemen, then, came to a peak in the mid-1760s because of the conjuncture of several circumstances. High prices and economic recession focused attention on the poor consumer. Nearly two decades of public abuse associated with the middlemen of the corn and meat trades made Englishmen aware

[162] John Bryant (ibid., March 1765).

of their entrenched position in the economy. When the old nostrums of Tudor and Stuart times were reintroduced, they not only failed, but actually worsened the existing problems of scarcity and high prices. The poor, conditioned to attribute food shortages to artificial rather than natural causes, hardly needed the encouragement they received to deal with a popular scapegoat that the authorities were unable or unwilling to regulate. The aristocratic and gentry leaders of rural society themselves were not unaffected by the lengthy denunciation of middlemen in the 1750s and 1760s. Many of the ruling orders may honestly have believed the food shortage was artificially contrived by the speculation of middlemen. There were however more cogent reasons why the ruling class played such an equivocal role in the extensive hunger riots of 1766. Prejudice against middlemen was a surface manifestation of underlying tensions which were affecting society after the mid-century.

3 The Role of the Authorities in the Provincial Hunger Riots of 1766

A crucial feature of the hunger riots of 1766 was the initial encouragement given to the mobs by the ruling orders in the countryside. In an age when prompt action by the local authorities invariably snuffed out riots before they became a real social threat, the restraint of the majority of the gentry-magistrates towards riotous mobs was extraordinary and tantamount to sanction. John Wesley, who was as experienced as any of his contemporaries in facing dangerous mobs, noted in his diary that 'Wherever a mob continues any time, all they do is to be imputed not so much to the rabble as to the Justices'.[1] In the critical early days of the rural riots when many of the dispossessed waited to see how the militant minority fared in their initial challenge to authority, the lesser gentry, who dominated the parish benches, stood aloof from the aristocratic county representatives of the national government on the one hand, and the commercial and industrial interests on the other, and scarcely troubled to conceal their sympathy for the riotous poor. But the magistrates not only refrained from effective measures to crush the initial disorders, they actually abetted other members of the landed and industrial interests in their encouragement of the people to regulate markets and reduce the prices of provisions by force. By this means, they

---

[1] *The Journal of the Reverend John Wesley, A.M.*, ed. Nehemiah Curnock, Standard Edition, 2nd ed. (London: Charles H. Kelly, 1781) v 250. See also Rudé, *The Crowd in History*, p. 262; Frank Ongley Darvall, *Popular Disturbances and Public Order in Regency England* (London: Oxford University Press, 1934) pp. 244–5; and Frederick Clare Mather, *Public Order in the Age of the Chartists* (Manchester: Manchester University Press, 1959) pp. 60–1.

diverted the rioters towards middlemen and large farmers, and away from the landed and industrial interests. Unlike other agrarian disorders of the century, the riots of 1766 did not involve direct attacks on landowners or manufacturers. Thus while not actually inciting the riots, the actions of the magistrates certainly gave them direction. Only belatedly, when the scale of disorder frightened them, did the gentry-magistrates close ranks with the aristocracy and other rural leaders to crush what they had come to fear was the start of social revolution.

I

That the magistrates were unusually derelict in their duty to suppress popular disorders in September 1766 is evident from the comments of their contemporaries and the course of events in urban and rural riots.

Understandably it was those Ministers responsible for domestic security who at this time most frequently complained of the negligence of the rural justices. Lord Barrington at the War Office wrote to the Earl of Shelburne, Secretary of State for the Southern Department, that troops sent 'to aid the civil power' had never been used, and commented that 'the magistrates at this juncture have less spirit or prudence than usual'.[2] Henry Conway, Secretary of State for the Northern Department, reflected the same opinion when he called upon the Duke of Marlborough, the Earl of Berkeley, and other Lords-Lieutenant to use their influence as great landowners to encourage the magistrates to act decisively against the rioters.[3] Yet it is evident from independent sources that Conway's conclusion that 'a want of activity in the use and exertion of the civil powers by the ordinary magistrates seems to be among the chief causes of these continued outrages' cannot be dismissed as a customary ploy to divert attention from the Ministry's own shortcomings.[4]

Private gentlemen, with no apparent vested interest in apportioning blame, wrote of the magistrates being 'backward to exert themselves'.[5] John Pitt, reporting conditions in the West Country

---

[2] Barrington to Shelburne, 2 October 1766, *Shelburne Papers,* vol. 132, fol. 13–14.

[3] Public Record Office, *Domestic Entry Book,* vol. 142, fol. 12–14.

[4] Ibid.

[5] James Harris to Hardwicke, 3 October 1766, Add. MSS., 35607, fol. 295.

to Lord Hardwicke, a former Lord Chancellor perennially concerned with the dangers of civil commotion, observed of the Gloucestershire riots that 'had the civil magistrates acted with the least degree of prudence all this trouble would have been saved the government and the lives of the poor creatures to the service of the country'.[6]

That more timely, vigorous action by the rural magistrates might well have stifled the riots is suggested by the situation in the towns where disorders in 1766 were quickly brought under control. In such urban centres the magistrates rapidly devised measures of self-help. When market disturbances threatened to spread throughout Gloucester, civic leaders immediately ordered all citizens off the streets and thereby pacified the city,[7] while at Norwich the mayor and his fellow magistrates armed the 'respectable citizens' with staves and dispersed the numerous rioters without military aid, during a weekend of violent protest.[8]

Certainly urban magistrates did not face the same problems as their rural counterparts. While city mobs could form and melt quickly away when peace forces arrived, only to reappear elsewhere, there was a narrower geographical limit upon their riotous activities than was the case for rural mobs. In urban centres there were always considerable numbers of readily identifiable 'respectable citizens' who could be trusted to fight to protect property in times of crisis, although at times 'wrong elements' occasionally received arms in the confusion of a large-scale riot. Such was the case in Norwich where rioters passed themselves off as respectable citizens and joined the armed *posse comitatus* before informers denounced them to the magistrates.[9] The problems posed at various times by the metropolitan mobs were peculiar to London and they are dealt with at length elsewhere. But even here the ready availability of troops to assist such competent magistrates as Sir John Fielding and Saunders Welsh mitigated the difficulties of controlling a vast slum population.[10] In contrast, there were

[6] John Pitt to Hardwicke, 18 December 1766, Add. MSS., 35607, fol. 335.
[7] John Pitt to Hardwicke, 29 September 1766, Add. MSS., 35607, fol. 290. *Gazetteer and New Daily Advertiser*, 8 October 1766.
[8] *Gazetteer and New Daily Advertiser*, 4 October 1766.
[9] Mayor of Norwich to Town Clerk, Norwich, 20 October 1766, Norwich City Record Office, Norwich, *Depositions*.
[10] Both these magistrates were valued by the poor and the Ministry, who paid them an annual salary.

To the P O O R,

THE Magistrates pity you, and you may be affured they will ufe every Endeavour to obtain PLENTY and CHEAP-NESS of PROVISIONS.

The greater Quantity which is brought to Market the more plentiful and cheaper it muft be.

But if the Country are driven out, and not fuffered to come in Peace, there can be neither Plenty nor Cheapnefs.

Rioting will ftop the Provifions from coming to Market, and will increafe the prefent Diftrefs of the Poor, and, at the fame Time, will make it impoffible for the Magiftrates to do any thing to ferve them.

In Compaffion to your Diftrefs the Magiftrates would not read the Proclamation, they wifh to avoid it.—For God's fake do not drive Things to Extremities: The Magiftrates are fworn to keep the Peace, and in all Events they muft do their Duty.

often dangerous delays before troops came to the aid of isolated rural magistrates. But commentators did not miss the lesson to be taken from the success of prompt suppression of the Norwich and Gloucester rioters. If further confirmation of the value of swift action against mobs be required, one need only observe, on the one hand, the effective actions taken by the local authorities against hunger rioters in 1756–7 in circumstances potentially much more dangerous than those of the summer of 1766, and the results of the belated co-operation between the magistrates, the army and Ministry, which occurred when the gentry became terrified by the extent of the disorders after the third week in September 1766, on the other.

Gentry-magistrates in 1766, however, did more than just tolerate the initial disorders. Some gentry encouraged the distressed poor to regulate markets for themselves and control the activities of middlemen, who were little impeded by the erratic enforcement of the old Tudor and Stuart statutes for the protection of consumers. Witnesses at Stroud in Gloucestershire reported that on 19 September rioters threatened to pull down Timothy Ratten's mill and declared that they had all the gentlemen on their side and that the Earl of Berkeley had given them three guineas.[11] One leader of the 'regulators' in Norwich market cried : 'Damn them, I have an order from the Gentlemen to serve them all alike and make no exception of none.'[12] Other gentry directed attention towards the larger farmers. Early in the food protests in Gloucestershire, the gentry of the county announced their intention of forcing the farmer to reduce the high price of grain. The implication of the gentry's attitudes and actions was that the food shortage was artificially contrived for their own profit by middlemen and wealthy farmers, whose interests were often seen by the landed interest to coincide.[13] John Pitt noted the dangers of such responses :

[11] *Treasury Solicitor's Papers*, T.S. 11/5956/Bx1128.

[12] Evidence against Brown, a leader of a riot in the market on 27 September 1766, Norwich City Record Office, Norwich, *Depositions and Case Papers*.

[13] A correspondent who urged the opening of the ports for importation to convince the public in January 1757 whether the scarcity was natural or artificial anticipated the opposition of factors and rich farmers (*Gentleman's Magazine*, xxvii [1757] 32).

The high price of provisions was felt by all with this difference that the rich received a reciprocal advantage in the advance of their incomes whilst the poor had no other recourse than in the advance of their wages. Their clamours raised to procure these were artfully turned to the cause of the other, nay foolishly by many who at the same time they were advantaged by the high sale of commodities grumbled to buy at an equal price.[14]

Pitt's estimation of the gentry's responsibility for the riots corroborates the rioters' frequent claims that 'the gentry are with us':

. . . the generality of the gentry in the country have formed a more erroneous judgment of the present distressed times than ever the mob, that that countenance which the mob has received from many has encouraged them to the lengths they have run and that many that are under sentence of death thought they were doing a meritorious act the very moment they were forfeiting their lives.[15]

Plainly the middling and lesser landowners, who dominated the local benches, were not the only influential members of rural society who were anxious to divert the attention of the poor towards middlemen and richer farmers. In the West Country where prices of food were high and employment short, clothiers wished to prevent a repetition of the agitation of cloth workers for higher wages and better working conditions which had led to serious disorders in the 1750s.[16] Employers, correspondents asserted, caused labourers discontented with scarce work and dear provisions to start the first riots 'by recommending a hint in the newspapers' and 'every alderman, common councilman, shoemaker and shopkeeper joined in this encouragement'.[17] Gentlemen-clothiers and others of the capitalist manufacturing interests sat on the bench alongside the landowners in the clothing counties and exhibited similarly equivocal responses to the food riots of 1766.

Inevitably the attitude of the rural leaders stimulated violence by increasing the exasperation of the dispossessed with the authors

---

[14] John Pitt to Lord Hardwicke, 20 December 1766, Add. MSS., 35607, fol. 340.

[15] Ibid.

[16] *Victoria County History, Wiltshire*, iv 64 *et seq.*

[17] John Pitt to Hardwicke, 21 December 1766, Add. MSS., 35607, fol. 341.

of an apparently artificial food shortage, and in the process focused the rioters' hostility on middlemen and larger farmers. In relatively unsophisticated, agrarian societies the ill-educated poor frequently attribute natural shortages of food to divine retribution and stoically endure them. When shortages appear artificial they direct their anger against those held responsible for their distress.[18] In September 1766, the authorities plainly indicated to the poor that middlemen and large farmers were creating an artificial shortage, when they promised to force farmers to lower prices, encouraged the enforced sale of provisions at 'just' prices and formed associations of gentlemen to prosecute forestallers, engrossers and regrators in conformity with the government's proclamation of the old paternal statutes.

To appreciate fully the unusual character of the rural magistrates' responses to the initial disorders of the late summer of 1766, one must take into account the traditional attitudes of the gentry-magistrates and landowners towards riots generally. Whig politicians, while not welcoming civil commotion, frequently regarded minor metropolitan disorders as relatively innocuous safety-valves for the surplus energies of the populace. An anonymous contributor to the *St. James's Chronicle* expressed this typical view :

My dear countrymen may proceed in their divisions and quarrels. They are never more amiable in my sight when they have rolled each other in the kennel. No music more sweet than the crashing of windows; and when they are of as many colours as a painted Indian, when they are beaten black and blue, and yellow, when their noses stream with blood, and they have not an eye to see with, why they are brave boys, hearts of oak, and bold Britons. And thus much is well in play. But let them proceed no further. Let them not insult the magistrate, nor obstruct the execution of the laws. Struggles against the laws are the convulsions of expiring liberty. If ever we are rebels, we shall soon become slaves.[19]

---

[18] Edmund Burke warned of the danger of implying that conditions of scarcity were not inevitable. He believed that dangerous riots were possible if the poor came to believe 'man's ingenuity could improve things'. He asserted that the government in 1767 had raised the hopes of the distressed poor without doing anything in the spring and summer of that year (*Parliamentary History of England [1765–1771]* [London: Longmans, 1813] xvi 390).

[19] *St. James's Chronicle*, 12–14 May 1768.

On occasion magistrates themselves headed mobs in an age of inadequate policing. Justices organised attacks on Papists, Dissenters and Methodists in defence of the established church or the social and political order in the eighteenth century.[20] In times of serious disturbance, the magistrates first formed associations of gentry who collected with their servants and suppressed the mobs. In doing this, they were in effect calling out the 'power of the county' as provided under Common Law. Often the rural leaders regarded the *posse comitatus* as scarcely more than a loyal mob. Thus the Marquis of Rockingham as Lord-Lieutenant of the West Riding urged that lieutenants of counties should not tamely yield up militia lists to angry mobs in 1757, and advocated the creation of a counterforce : '. . . they should arm the townspeople and have a strong mob in readiness to oppose to any mob which should dare to attack them.'[21] The preference of the gentry and aristocracy for 'loyal mobs' rather than more efficient army units reflected the traditional suspicion of standing armies. Many of the ruling orders saw their ability to raise mobs as an effective counterbalance to the threat of military despotism. Most were willing to risk the occasional dangers and inconveniences of too little police supervision of the populace in order to prevent a return to Stuart or Cromwellian centralism or the establishment of a Bourbon style of despotism. Publicly men like the Duke of Newcastle declared their affection for the mob, and acknowledged their debt to it : 'I love a mob. I headed a mob once myself. We owe the Hanoverian succession to a mob.'[22] To such men, the opportunity to raise a mob was the last barrier to protect property and privilege. The care of the War Office always to make clear the subordination of the troops to civil authority, however, reassured many who feared too great dependence on the army.[23] The growing scale of social protests in the late eighteenth century finally dispelled any lingering doubts about the use of military force and encouraged the acceptance of a regular police system.[24]

[20] Thompson, *The Making of the English Working Class*, p. 74.

[21] Rockingham to Newcastle, September 1757, Rockingham MSS., R1–105.

[22] *Westminster Journal and London Political Miscellany*, 11 June 1768.

[23] See instructions to army commanders to assist magistrates suppress disorders (*Marching Orders*, WO5–54, *passim*).

[24] 'It is certainly necessary to encourage the Civil Magistrates, and support this Authority; for if that is not done, we must either be governed by a mad, lawless mob, or the peace be preserved, only by military force' (Newcastle to Rockingham, 13 May 1768, Rockingham MSS., R1–1052). See also Newcastle to Mansfield, 13 May 1768, Add. MSS., 32990.

If the gentry and aristocracy were somewhat indulgent towards metropolitan political rioters before the 1760s, and on occasions organised 'bully boys' to break up Methodist open-air meetings in rural parishes, they were more sensitive to social protests in the countryside. Few landowners remained indifferent when serious disorders approached their estates. Unrest in the countryside threatened all they prized most. The distinctly labouring-class character of the provincial rioters had levelling implications for the privileged landed interest. Might not a mob of landless labourers readily turn from regulating markets or destroying houses of industry to correct more profound economic and social anomalies? In a century when industrial workers were in scattered pockets throughout rural England, gentry recognised the dangers of class antagonisms polarised over industrial disputes. However pleased they might have been by the embarrassment of the socially ambitious industrialists, magistrates promptly suppressed riotous workers.

While paternalism dominated the feelings of the landowners towards the agrarian poor, and the magistrates reflected in their actions in times of distress the attitudes of the landed interest, the relationship between the poor and the landowners always presumed a proper deference of the one towards the other. The gentry found disquieting such levelling cries as those of the protesters against the new houses of industry of East Anglia in 1765–6 that the ground was as free for them as for the magistrates.[25] Witnesses always noted such unwelcome assertions and duly reported them to the authorities, whether they were uttered by poor-law demonstrators, hunger rioters, coalheavers or Wilkite mobs. Thus the rural magistrates were well aware that once disturbances developed, rioters had a most inconvenient habit of remembering a variety of long-standing grievances. The ultimate crime, to the privileged landed interests, was to threaten the social order, for they still subscribed to the creed of their Tudor ancestors : '. . . Take but degree away, / Untune that string, and hark what discord follows.'[26]

Social considerations are evident, too, in the response of the middle and upper classes to the government's handling of riots.

---

[25] *Marching Orders*, WO5–54, pp. 53–8 and *passim*.
[26] Shakespeare, *Troilus and Cressida*, I iii 109.

Political factions made little attempt to create capital from the government's methods of suppressing hunger and industrial riots in the 1760s. Only George Grenville, still smarting from the loss of office and the rejection of his economic policy for America, sought to censure the Chatham–Grafton Ministry in November 1766.[27] Even then his attack was upon their lack of foresight and the illegality of their suspension of grain exports by Order-in-Council, rather than upon the employment of excessive force. Similarly there was little political response to the government's handling of the several pre-industrial riots of the weavers, seamen, coalheavers and others in the spring and summer of 1768. Far different was the denunciation of military brutality which followed the political riots centred around the person and causes of John Wilkes, later in the same year. The different attitude of the authorities themselves towards agrarian and pre-industrial riots on the one hand, and political disorders on the other, is apparent from a study of the sentences given the respective sets of rioters: capital punishment and lengthy prison sentences for the one, compared with short terms of imprisonment or fines for the other.[28] Military suppression of riots with obvious class implications might be severe without public outcry in the press or Parliament; but political disorders which crossed class lines generated bitter criticism of military brutality. Apparently the articulate upper and middling sorts distinguished between two categories of riot : rural hunger and urban industrial riots which had obvious class connotations, and political riots which did not. (This of course does not deny that especially in the metropolis the distinctions between categories were often blurred.) In the case of the provincial hunger and metropolitan pre-industrial riots most gentry could agree with the pamphleteer who noted : 'Distress furnishes an apology for violence; the levelling principle begins to operate, and the chain is broken which connects the higher ranks with the lowest. Subordination is lost, and he [the labourer] regards the landlord and farmer as oppressive tyrants.'[29]

---

[27] Mr Grenville to Earl Temple, 18 November 1766, *Grenville Papers*, III 341–3.

[28] Rudé, *Wilkes and Liberty*, Appendices III–V.

[29] *Considerations on the Exportation of Corn*, p. 56.

Rural magistrates were with reason sensitive to threats of large-scale disorders in an age when violence was a normal ingredient of daily life and no effective police force existed. Most cases heard by justices concerned common assault. Prevention of general anarchy in times of distress required of local authorities considerable courage and intelligent foresight. Magistrates were greatly concerned about social and economic conditions which created unusual stress in rural society. They were often willing to forego their own immediate economic advantage when the prospect of a famine required it.[30] Magistrates normally attempted to anticipate stress-causing situations.

But by the 1760s there was a great deal of confusion and uncertainty among the rural magistrates about the efficacy of much of the paternal legislation such as the statutes dealing with wages, apprenticeship and the variety of offences by engrossing middlemen and farmers which were designed in simpler times to preserve the poor from economic disaster.[31] In the first half of the century food prices were low and the conditions of the poor were improving. The restrictions on forestalling, engrossing and regrating had ceased to be enforced except in rare crisis years. In this period middlemen entrenched themselves in the increasingly sophisticated economy. Only after the mid-century did attention centre on their functions and pressure grow for the reimposition of controls. But the fear of loss of revenue to more liberal, rival markets ensured that clerks of markets continued to ignore the old restrictions against middlemen, despite their mounting unpopularity.[32]

After 1757 magistrates across the country followed the example of their brother magistrates in Gloucestershire and refused to fix the wages of the industrious poor.[33] The practice of setting the assize of bread in rural England was also in the process of abandonment, although the pace of change was slower than was the

---

[30] Rose, 'Eighteenth Century Price Riots and Public Policy in England'.
[31] See Chapter 2 above.
[32] R. Wright, Town Clerk of Warwick, to the Rt Hon. the Earl of Abercorn, 9 March 1765, *Committee on High Prices of Provisions*.
[33] Christopher Hill, *Reformation to Industrial Revolution; The Making of Modern English Society, 1530–1780* (New York: Random House, 1967) I 220.

case for the system of fixing wages, and bread assizes particularly in urban centres continued into the next century.[34]

Depression of industry at a time of high food prices soon after the Seven Years' War on the one hand, and the practice of *laissez-faire* towards the poor, increased the numbers thrown on to relief and strained parish resources. In such areas as East Anglia the magistrates sought to rationalise the system of relief and reduce costs by establishing houses of industry to shelter the poor of several parishes. The immediate result of attempting to send the poor from their home parishes to obtain indoor relief was increased unrest among the indigent who claimed the right to support in their native parishes, and riotous attacks on the institutions.

Despite the trend away from paternalism, magistrates in times of severe economic crisis reverted to regulations appropriate to a less sophisticated economy. They were slow to accept the increasingly popular theory that the 'necessaries of the poor' should form merely another system of trade in which the laws of supply and demand operated, and unlimited profit-taking was acceptable.[35] The principles of the older 'moral economy' and the medieval concept of the 'just' price lingered in the minds of both the magistrates and the poor.[36]

[34] Country magistrates rarely set the assize at all. It was assumed rural households baked their own bread. The Webbs cite a correspondent of the *London Chronicle* (24 December 1761) who petitioned the House 'in the name of the many thousand rural housekeepers in England . . . for some law respecting the bakers in the country, who are now almost unregulated' (S. and B. Webb, *The Assize of Bread*). The assize of bread was a 'source of constant friction in the eighteenth century. It was abandoned in the metropolis in 1815 and in the rest of the country in 1836.' The chief problems were threefold : (1) Bakers did not purchase on the same basis across the country, e.g., London bakers bought flour not corn, other bakers bought corn and had it ground. Prices of flour did not vary directly with those of corn. (2) Corn and flour prices fluctuated, making it difficult for magistrates to keep the assize up to date. (3) Errors in assize prices could seriously affect supplies, e.g., too low prices would discourage imports, say, into London (*Observations and Examples to Assist Magistrates in Setting the Assize of Bread Made of Wheat under the Statute of the 31st George II* [anonymous pamphlet, London, 1759] pp. viii *et seq.*

[35] Sir John Fielding to Lord Abercorn, 5 February 1765, *Committee on High Prices of Provisions.*

[36] Thompson, p. 68 and *passim.* R. B. Rose (op. cit.) relates the price-fixing riot to the medieval doctrine of a 'just price'. See also E. P. Thompson, 'The Moral Economy of the English Crowd in the Eighteenth Century', *Past and Present,* no. 50 (February 1971) pp. 76–136.

The natural interests of their privileged social and political position dictated that gentry-magistrates be vigilant against all disorders. The threat of presentment before the judges of assize for neglect of their duties was hardly necessary to persuade them to honour their obligations for rural peace. Fear of revenge at the hands of the rioters occasionally inhibited the conduct of the more timid, isolated justices, who were acutely aware of their vulnerability to cattle-maiming, rick-burning and even assaults on their homes and persons. But there were few if any occasions when rural benches petitioned the Lord Chancellor for the removal from the Commission of the Peace of one of their brothers on the grounds of negligence. Self-interest, blended with considerations of social prestige which accompanied the appointment to the county bench, was sufficient incentive for the overwhelming majority to perform their duties adequately where other pressures failed.

The traditional attitude of gentry-magistrates towards rural disorders, then, may be summarised as one of extreme sensitivity. Because they feared such threats to the social structure, they acted to avert the conditions likely to stimulate violent protests.[37] Associations of gentlemen provided subsidised grain for the poor; petitioned Parliament for control of grain exports, grain distilling and starch making;[38] and prosecuted individuals for contravention of the statutes against engrossing, forestalling and regrating. When, despite their efforts, outbreaks occurred, magistrates moved promptly to suppress them, either with local forces or with regular troops. Why, then, did the local authorities in rural areas go against all precedent in the late summer of 1766, and not only permit disorders to develop, but actually encourage the populace to take the law into their own hands to regulate markets and enforce 'just' prices?

II

One possible explanation for the equivocal response of the gentry is that they generally misunderstood the nature of the food

[37] The 1760s saw the last sustained effort by rural authorities to apply the outmoded regulations of the Tudor and Stuart era in the face of economic reality.

[38] The distilling and starch-making industries were particularly provocative to the poor in times of scarcity because they produced luxury goods at the expense of the food supply.

shortage and genuinely believed it was due to the actions of monopolising farmers and speculating grain dealers. Certainly their unsympathetic attitude towards these interests was consistent with public antipathy towards large farmers and middlemen which was very evident after the mid-century. Both the newspapers and the actions of the government encouraged the view that the developing crisis was due to an artificial rather than a natural shortage of grain. In July and August 1766 press estimates of the harvest prospects were mixed, but the majority promised heavy crops, which later encouraged the popular belief that the poor were starving in the midst of plenty. The action of the Ministry in proclaiming the old statutes against forestalling, engrossing and regrating suggested strongly that the culprits were the larger farmers and middlemen. Continued heavy shipments of grain to the ports after the ending of the six months' embargo on grain exports encouraged the view that these two interests were making high profits at the cost of draining the country of grain. Rumours transmitted through the market places distorted the truth further and caused the poor to panic at the prospect of an outright famine. In such circumstances, the gentry must have found it easy to associate themselves with the poor in their resentment of the larger farmers and middlemen of the provisions trade.

But country gentlemen were not as easily misled about the true quality of the growing crops as were the 'spruce Londoners' who reported on their journeys through the countryside and estimated harvest yields without, critics said, being able to distinguish between wheat and barley.[39] Country-bred landowners were quite able to recognise that the larger ears of growing corn were the inevitable result of a hot August following hard upon a very wet, early growing season, and that they promised a lightweight, coarse grain in a year when a better than average crop was essential if a food crisis was to be averted. Almost certainly even those gentry living outside the heavy corn-growing areas of southern England learnt of the true prospects of the harvest through the normal channels of communication of their class: the race meeting, the hunt ball or some other county function. Such accurate information would eventually have reached even the lesser parish gentry. It is unlikely that many of the gentry in September 1766 thought that the food shortage was artificial.

[39] *Gentleman's Magazine*, xxviii (1758) 565.

Probably many magistrates realised the danger that some un-scrupulous businessmen might profit from a natural shortage and in the process aggravate the existing crisis, and they therefore attempted to enforce rigorously the practically defunct anti-middlemen statutes. But such moderate and provident actions fell far short of the partisanship most of the county leaders displayed towards the large farmers and middlemen. A more convincing explanation must be sought beyond the circumstances of the riots themselves. It recognises not only the social tensions with'n the landed interest itself, but the effect on the rural leaders of an outburst of class conflict which occurred in 1756–7.

Some knowledge of the experiences of the gentry-magistrates less than a decade earlier, during another period of popular unrest, is essential to the understanding of their actions in 1766. Serious disorders centred around the implementation of a new Militia Act in 1757. Disturbances were most dangerous in the eastern half of England, particularly in the vicinity of the Humber. Although the most disaffected region in 1757, the East Riding, was calm in 1766, several counties such as Lincolnshire, Nottinghamshire, Derbyshire, Hertfordshire, Northamptonshire and Norfolk experienced serious riots in both years.[40] Even in counties where the farmers and their labourers did not violently protest against the Militia Act, such as Gloucestershire, the rural leaders were well aware of the threat of serious disturbances and took pains to advertise in local newspapers to quieten popular resentment: 'The Good People of England' had misunderstood the Militia Act, none were to be forced to travel more than six miles for exercise, none were to be sent out of the county unless there was imminent danger of invasion or open rebellion, none would go out of the kingdom or serve longer than three years, during which time they would be exempt from statute work, service as peace officers or in the army; after actual service they were to be entitled to set up in trade.[41] The *Gloucester Journal* quoted from an 'Admonition to the Militiamen of Norfolk' which promised that personal hardships caused by militia service would be relieved by the justices of the peace.[42]

The major concern of the poor in 1757 was enforced service

---

[40] Western, *The English Militia in the Eighteenth Century*, pp. 291–4.
[41] *Gloucester Journal*, 6 September 1757.
[42] Ibid., 8 November 1757.

abroad. Although the militia was traditionally a territorial defence force, the authorities had forced the Somerset and Dorset battalions to embark for America in the previous year.[43] This action had caused widespread resentment among the poor, and they anticipated an extension of compulsory recruitment for overseas service in 1757 as a result of the new Militia Act. A common cry of the rioters was 'better to be hanged at home than scalped in America'.

Less emotional but scarcely less serious objections concerned the pay of militiamen on active service, and the support of their dependents. The hostility of the poor was all the more menacing to the ruling classes when it was complemented by middle-class anger at growing parish expenditure, the weight of which fell disproportionately on the shoulders of the rural middling sort. 'Which of you bunting-ars'd coated fellows will maintain his [labourer's] family?' asked one contemporary. J. R. Western, the historian of the eighteenth-century English militia, notes that one letter of protest to the Lincolnshire magistrates reads like a 'middle-class manifesto' : 'If the Just-asses and the other start up officers that buys a commission for a trifle and sells his Nation to make his fortune when he comes abroad, and throws thousands of poor men's lives away about it, such men as those sho'd behave well to their tenants at home. Then they would have the countreys goodwill, for 'tis the farmers that maintains both the poor and such as they too. . . .'[44] The new Militia Act angered the farmers because it threatened heavier parish rates at a time when high food prices increased the number of poor on relief and farm income was down without any corresponding reduction in rents.[45]

In 1757 the new Militia Act stirred bitter class feelings in the countryside. Mobs of labourers led by farmers attacked both gentry and peers as they sought to destroy the militia lists. At Buckrose, in the East Riding of Yorkshire, a large body of farmers and country people 'out of forty townships in the Wapentake of Buckrose . . . arm'd with guns, scythes, and clubs rose on account of the militia act' which they claimed 'was a great hardship upon the country, by compelling the poorer sort of people

[43] Western, p. 298.
[44] Ibid., p. 300.
[45] Ibid., p. 299.

to contribute equally with the rich. . . .'[46] Another typical incident occurred at Mansfield, Nottinghamshire, where a mob of five hundred took lists from an assembly of gentlemen and 'none of the gentlemen . . . escaped without receiving marks of their resentment'.[47] One mob broke Lord Vere Bertie's windows and planned to go to the Lincoln races to attack the nobility, whom they blamed for the Militia Act. A similar incident occurred at Northampton.[48] The unusual combination of farmers and agricultural labourers against the landed interest presented a dangerous social threat to the privileged leaders of rural society in 1757. Dark murmurings that the gentry had enjoyed their broad acres for too long indicated the tendency of riots precipitated by specific grievances to expand to include larger social issues. Isolated on their country estates, the landowners felt the hot breath of social revolution.

Food scarcity and high prices created a background of discontent against which the riots over the Militia Act occurred. Riots, precipitated by the formulation of lists of men to serve in the militia, often spilt over into attempts to regulate markets by force. The authorities in Yorkshire proposed restrictions on grain exports to placate mobs protesting against the Militia Act.[49] The connection of militia and food riots in 1757 had important implications less than a decade later. Sporadic riots against the militia regulations continued into the 1760s, notably in Buckinghamshire and Northumberland.[50] It is reasonable to suppose that in 1766, when protests against high food prices began once more, the landed interest feared a resumption of events surrounding the militia riots of 1756–7.

When the hunger riots broke out in September 1766 the gentry were determined to avoid the social isolation in the face of a hostile combination of farmers and labourers that had terrified them a decade earlier. They now cultivated their traditional identification with the poor against the corn dealer and large farmer. They mistakenly believed, against all their experience,

---

[46] *Gentleman's Magazine*, xxvii (1757) 431.
[47] *Gloucester Journal*, 20 September 1757.
[48] *Gentleman's Magazine*, xxvii (1757) 430.
[49] Add. MSS., 32875, fol. 285–6, 411. Western (p. 300) cites various evidence to support the connection of food and militia riots in 1757.
[50] *Calendar of Home Office Papers* (1766–9) no. 1230.

that disorders could be limited to mild measures to regulate markets and enforce 'just' prices.

The ruling classes were divided. The lesser gentry with their smaller, less diversified holdings felt no common cause with the great landowners, who as Lords-Lieutenant represented the central government. They resented the failings of the national government to provide stable economic and political conditions favourable to social order. Such feelings reinforced the usual parochial suspicions of metropolitan interference. In the 1760s the parish gentry felt themselves in a reeling world. The fixing of wages by traditional methods was no longer operable. Following serious disturbances among West Country weavers in 1756–7, the magistrates of Somerset and Wiltshire refused to set the wage assize, and thereafter other magistrates followed suit.[51] Machinery for the establishing of price thresholds for corn bounty payments and the prices of bread was a failure.[52] The effects of war service were evident in the increased taxes and discontents of the veterans of the Seven Years' War. In this world of confusion, many rural magistrates blindly clung to old nostrums like the anti-middlemen statutes of Tudor and Stuart times. When these not only failed, but aggravated the very crisis they were supposed to alleviate, for a critical period the rural magistrates abdicated authority to the mob.

Perhaps at no time in the century, even in the early years when, according to the Marquis of Rockingham, probably more than half of the population were Jacobite in sympathy, was there greater danger of general insurrection.[53] Because of the disunity of the ruling classes and their isolation from important segments of the industrial, agricultural and commercial middle classes, the danger to the social order for a brief period was perhaps greater than in the prolonged period of Chartist and Luddite disturbances of the next century. The absence of an effective civil police and rapid communications compounded the dangers of insurrection in the eighteenth century. The widespread character of the agrarian riots of 1766 in the most populous belt of the English countryside posed problems for the divided authorities that nearly proved insurmountable.

---

[51] De L. Mann, 'Textile Industries Since 1550'.
[52] S. and B. Webb, 'The Assize of Bread'.
[53] Rockingham to Newcastle, 17 May 1768, Add. MSS., 32990, fol. 83–6.

Because they directed the social resentment of the poor towards the apparent authors of an artificial shortage of provisions—food jobbers, corn factors, bakers and large farmers—the gentry must carry a major share of the responsibility for the spread of serious hunger riots in the late summer and autumn of 1766. Their technique of diverting attention from themselves was successful and there appear to be no records of attacks on gentry or peers by hunger rioters in 1766–7. The potential for such attacks is apparent from the threats carried in a letter forwarded to Lord Shelburne:

The continual dearness of provisions obliges us to lay some proposal before you:— On the first assembly of the Mobb the worthy gent[n] of this county met and declared in public they would use some means to oblige the farmers to reduce the high price of grain & we find all is dropt and nothing done for us our extreem necessity desires youl please to refer to the clothiers in & about Dursley and Painswick this serves to advise it is agreed between a set of men that may be depended on who have taken a list of the most substantial farmer tenants belonging to you and most Gentlemen of the county unless the prices are reduced immediately we are determined and certainly will take revenge by firing their houses, barns, stacks of corn etc if you are willing to prevent this dreadful experiment lay some injunction on your tenants or by all thats good it shall be put into execution.[54]

On other occasions earlier in the century it was not uncommon for hunger rioters to attack gentry. This was the case for example at Newcastle in July 1740.[55] Paradoxically at the very time the actions of the local authorities stimulated widespread violence which eventually appeared to threaten the social order, they ensured that a more dangerous coalition of the rural middlemen and middle-class farmers with the lower-class labourers against the landed and industrial interests could not occur. The landed interest acted in this fashion more from instincts of self-preservation than from any coolly-conceived plan. They temporarily forgot their customary concerns for the dangers inherent in any rural disorder, until they suddenly awoke to the extent of the riots, and terror at the prospect of revenge at the hands of the dispossessed drove them to co-operate with the Ministry, the large

[54] *State Papers*, SP 37/6, fol. 7–15.
[55] *Gentleman's Magazine*, x (1740) 355.

aristocratic landowners, the rural middle classes and the army to restore order.

III

To a considerable degree the actions and attitudes of the Ministry in August and September 1766 affected the character of the local authorities' responses to the hunger riots. The government's failure to renew the embargo on the export of grains when it expired on 26 August or to permit the continued importation of duty-free American grain aggravated the food crisis, and the proclamation of the old Tudor and Stuart statutes against forestalling, engrossing and regrating provided the magistrates and the poor with credible scapegoats for the food crisis. The reimposition of the export embargo on 26 September came too late to avert serious disturbances. At the same time as their action of proclaiming the old statutes provoked riotous attacks on middlemen and large grain farmers, who were suspected of conspiring to denude the country of grain to make large profits, the Ministry neglected to take elementary precautions to crush disorders. By omitting to place the army in southern England on alert, the War Office left the most effective units of riot control, the cavalry, with their horses at grass in distant pastures. Robbed of their manoeuvrability, the cavalry frequently in the first weeks of rioting had to march dismounted through the hot, dusty lanes to the disaffected areas, a circumstance not conducive to the good-humoured dispersal of rioters.[56] The Ministry's failure to give the rural governing classes leadership left the forces of order weak and confused. Gentry and magistrates for a while openly connived at the lawless acts of the mobs; the central authorities took over three weeks to absorb the significance of the mounting violence; and the delay in preserving the stocks of food and taking resolute action against the rioters ensured that when the inevitable suppression followed it would be severe.

Ministerial actions and attitudes to the agrarian disorders must be examined in the light of certain economic, social and political problems, some of which were inherited from earlier governments. The Chatham Ministry had barely entered upon its 'new freehold' in July 1766 when it faced a food crisis. As noted in

[56] *Marching Orders*, WO5–54, p. 305 and *passim*.

Chapter 1 above, the latest of several severe fluctuations in the prices of provisions, which had distressed the poor since the end of the Seven Years' War, touched off scattered food riots in Berkshire and the West Country. Prolonged bad weather and the expectation of renewed exports of grain caused a panic fear of outright famine to sweep through the markets and bring the poor to demonstrate their concern. The government gained a respite with the ending of the July rains. A dry, sunny August raised the hopes of the populace for a good harvest and lower food prices followed.

In the summer of 1765, Parliament had passed legislation enabling the Privy Council to suspend grain shipments abroad should an emergency arise during the summer recess.[57] Neither the Chatham Ministry nor the Rockingham Ministry provided for such exigencies in 1766. In part this omission was due to confusion about the duration of the emergency powers granted to the Council by Parliament in 1765.[58] But a more important reason seems to have been a failure to anticipate the food crisis which occurred in September 1766.

To assume that any Ministry in the 1760s had the means to collect and correlate all the essential economic information necessary to determine the true nature of the food crisis in September 1766 would be unreasonable. In fact the information reaching London was usually fragmentary, contradictory and out of date. Land communications were still poor, and the bureaucracy's ability to collect and interpret statistics was low. Quantities of grain exported in a given period, for example, were not known until the customs officers in London and the outports provided the figures. No systematic method of compiling tables of prices in the various rural markets was in force, although some journals attempted such a project without success.[59] Observers exaggerated the extent of disturbances. Ministers like Shelburne, Barrington and Jenkinson had their own sources of information, but these

[57] 5 George III, cap. 32.
[58] Harcourt to Jenkinson, 16 September 1766, Add. MSS., 38340. See also Hardwicke to Rockingham, 6 December 1766, Rockingham MSS., R1–722; Lord Shelburne to the King, 2 September 1766, *George III, King of Great Britain, 1738–1820: The Correspondence of King George the Third*, ed. Sir John William Fortescue, vol. I: 1760–7, 1st ed., new impression (London: Cass, 1967) no. 384, p. 391.
[59] Notably the *Gentleman's Magazine*.

were only incomplete, impressionistic surveys of conditions in different areas of the countryside. Newspaper reports were even less helpful. They were inaccurately based on the reports of volunteer correspondents, who drew heavily upon rumour. The government faced a populace even more dependent upon market rumours than they were themselves. In such circumstances it behoved the Ministry to avoid hasty measures which might only aggravate existing problems.

While the estimates of the harvest prospects were contradictory, the optimism of the majority supported the government in taking no special emergency measures during late July and August. Buckinghamshire, Bedfordshire, Oxfordshire, Warwickshire, Staffordshire and Worcestershire all looked for the finest crops of hay and corn ever known, despite the heavy rains.[60] Huntingdonshire expected a record harvest, although it would be delayed owing to the wet season.[61] Dulwich, Camberwell and Peckham reported a 'fair and more plentiful crop never known'.[62] One observer claimed the finest crops ever remembered in the western counties, although harvest hands were short.[63] As late as 8 September, a gentleman lately returned from the seven western counties confirmed the excellence of the crops there.[64]

Wrong assessments of the crops were partly due to an ignorance of farming on the part of London observers, but the tendency towards optimistic prediction was almost inherent in the speculative character of the corn trade. Rumours of shortages or abundance greatly affected prices. Newspapers tended to anticipate good crops most frequently out of a conscious or unconscious desire to maintain price stability. Some favoured districts however did enjoy good harvests where the early frost and summer rains had not struck; but frequently correspondents represented these as typical of whole counties. One observer for example wrote that crops were heavy in High Suffolk, although on an average crops in Suffolk proved disappointing.[65]

Taken at their face value, such press reports could well have misled the Ministry into thinking that the crisis might ease when

[60] *Public Advertiser*, 11 July 1766.
[61] Ibid., 30 July 1766.
[62] Ibid., 9 August 1766.
[63] Ibid., 18 August 1766.
[64] *Gazetteer and New Daily Advertiser*, 8 September 1766.
[65] *Public Advertiser*, 27 September 1766.

the harvest ended. Yet elsewhere in the press, warnings appeared. Correspondents reported severe flood damage to crops and stock in Gloucestershire, Oxfordshire, Worcestershire, Berkshire, Wiltshire and elsewhere.[66] Evidently a normal harvest would not be sufficient to replenish old stocks almost exhausted by the early summer; an extraordinarily heavy crop was essential to avert widespread distress. The demands for the government to renew the embargo on grain exports became more strident during August.

The net effect of the press was to confuse the picture entirely. One could substantiate any opinion on the harvest prospects from the 'facts' that the papers published in August 1766. More important than this confusion, the press distorted the true character of the food shortage, after the prices rose steeply in September. Having anticipated bumper harvests publicly, the publishers later, when faced with a shortage of provisions, had to choose between admitting their errors or declaring the shortage was artificial rather than natural. The hints they conveyed to the populace encouraged the riotous attacks on the middlemen and large farmers. The newspapers' conclusions that the shortage was artificial certainly emboldened the Ministry to move against the three deadly sins of forestalling, engrossing and regrating to which larger farmers and dealers in provisions were prone.

Even had the prospect of a food crisis appeared certain in early August, it is doubtful if the leaders of the new Ministry could have spared it much thought. Their preoccupation was with the question of survival – how to widen their narrow power base in the Commons by persuading one or more of the factions to join them despite Chatham's stubborn refusal to make political bargains.[67] The vulnerability of the new Ministry to pressure groups partly explains their failure to renew the export embargo on grain when it expired on 26 August. With reason, they believed that too early an intervention in the grain trade would have antagonised the independent gentry in the House. Although all landowners did not benefit equally from the export bounty system – indeed many suffered because while they grew little corn they were subsidising those who did through taxes – many regarded

[66] *Gazetteer and New Daily Advertiser*, 26 July and 29 August 1766. *Public Advertiser*, 30 July, 4 August and 12 August 1766.
[67] Brooke, *The Chatham Administration, 1766–1768*, p. 4 *et seq.*

the Corn Laws as symbolic of the supremacy of the landed interest, and therefore not to be meddled with lightly. Some unquestioningly accepted the Corn Laws as a necessary feature of the mercantile system. While the Ministry themselves were great landowners and not indifferent to the interests of their class, they had a wider view of the needs of the entire nation.[68] Thus they were willing to forego their own interests for the greater good in times of crisis. But politically it was always expedient that embargoes on grain exports be seen as a last resort. Pelham, Newcastle and West at various times reflected the attitudes of most landowners towards export embargoes : they should always be reluctantly imposed and of short duration.[69] It was necessary in early September 1766 to postpone action to stop the draining of the country's grain so that a familiar procedure could be followed, the first step of which was to proclaim the enforcement of the old statutes against forestalling, engrossing and regrating.

There were other pressing reasons why the government delayed the renewal of the embargo on grain exports. National grain policies must be set in the wider context of the country's total trade. During and after the Seven Years' War, there had been occurring a significant realignment of trade. Normal commercial relations with America had resumed with the repeal of the Stamp Act in March 1766, but the concern for trade expansion that the non-importation policies of the colonists had brought to the fore continued. That the English population had outstripped the nation's ability to produce food except in extraordinary seasons was not yet apparent, and many Englishmen continued to think of their country as the granary of Europe. With the urgent demands of famine-stricken Europe coming hard upon the heels of a six months' embargo on the export of grains, the government was seriously concerned about losing traditional markets. Because a healthy agriculture built upon the corn trade with Europe was the cornerstone of the economy, in the view of many of the ruling class, a delay in reimposing the embargo on grain exports was attractive to the Ministry. During the period of one month, grain

[68] Rose agrees with D. G. Barnes (*History of the English Corn Laws*, p. 16) that the government was sensitive to public opinion expressed in grain riots, which he believes were instrumental in reducing the effect of the highly protectionist Corn Laws (Rose, op. cit.).

[69] J. Harris to Hardwicke, 3 October 1766, Add. MSS., 35607, fol. 295. Newcastle to Mr White, 17 November 1766, Add. MSS., 32977, fol. 403–4.

left England at a rapid rate and compounded the problems of an existing shortage (between, that is, 26 August and 26 September).

Legally the Ministry was on unsure footing when it finally suspended grain exports by a proclamation of the Privy Council.[70] As Horace Walpole noted, such extra-Parliamentary action was not known before in peacetime.[71] That this was a cause of concern to the Ministers is evident from their correspondence. Yet it is not possible to explain the government's reluctance to reimpose the embargo on grain exports solely by reference to a preoccupation with legality. Had they wished, the government could have called Parliament together early in September when the first disorders occurred, or at least permitted it to meet on 10 September, as originally planned. In fact they postponed the session for sixty-two days.[72] Walpole explained their reluctance to convene the House by suggesting that the country would have been drained of grain in the interval between the issuing of the writs and the passage of legislation to cut off grain exports by the profit-hungry middlemen and large farmers.[73] But the rate of export was already so high that any increase in such circumstances would have been marginal. A more credible reason is that the natural leaders of society in rural England were needed most on their estates in times of general disorder. A further consideration was that when Parliament was in session during times of distress, Westminster became the focus of protest for the London populace. Probably, too, the Ministry welcomed a respite to organise a wider base of Parliamentary support.

More immediate in its impact than the delay in reimposing the embargo on grain exports was the Privy Council's proclamation of the enforcement of the old statutes against forestalling, engrossing and regrating on 10 September. This action achieved little more than a focusing of public attention upon a popular scapegoat, the middleman. The failure of the government to control the activities of the middlemen whom they publicly blamed for the scarcity of food was an open invitation to the poor to take

[70] The King to Shelburne, 23 September 1766, *Correspondence of King George the Third*, ed. Fortescue, p. 397.
[71] H. Walpole to Sir Horace Mann, 25 September 1766, Walpole's *Letters*, ed. Toynbee, VII 42.
[72] E. Langton to Hardwicke, 11 November 1766, Add. MSS., 35607, fol. 330–1.
[73] Walpole, *Memoirs of the Reign of King George the Third*, p. 263.

matters into their own hands and regulate markets for themselves.

More significantly, the actions of the Ministry encouraged the local authorities in the important role they played in the disorders. Gentry-landowners and industrialists found it convenient to accept the broad hint offered by the central government. Magistrates at first took an indulgent view of the riotous behaviour of the poor, and even discreetly encouraged it. They neglected to call for the military assistance that Barrington at the War Office was so anxious to provide until they suddenly became aware of the extent of the disaffection. A torrent of appeals from isolated estates and market towns then poured into the War Office, until with considerable difficulty the army crushed the rioters.

The Ministry's errors were due primarily to ignorance of the operation of the economy and lack of foresight. There had been indications of trouble for some time. Food prices had been high since 1764, and Parliament, alarmed at rumblings of discontent among the rural and urban poor, had held a number of inquiries into food prices in 1764, 1765 and 1766. By attributing the responsibility for the difficulties of the poor to middlemen of the grain and meat trades, together with larger farmers, and proposing no remedial action other than the more rigorous enforcement of the old paternal statutes in the interests of the consumer, these committees had exacerbated the problem.

When the central authorities finally became aware of the seriousness of the outbreaks in the fourth week of September 1766 they put pressure on the local authorities to utilise the military for the vigorous suppression of the rioters. Orders forbidding the distilling of grain liquors and starch-making followed the belated embargo on grain exports of 26 September. By the end of October most of the countryside was pacified. The work of arresting offenders extended over several weeks, during which time the army co-operated with the local magistrates. The Ministry then appointed special commissions in the worst disaffected counties to make examples of the rioters. Punishment was severe. The convicted felons were sentenced to hanging, transportation or service in the Royal Navy. While desultory protests continued into 1767, the authorities blackmailed dissidents by threatening to carry out suspended capital sentences if disturbances continued.[74]

[74] Dean Tucker to Shelburne, 11 January 1767, *State Papers*, SP 37/6, fol. 10, p. 155.

While many of the Ministry's errors were due to ignorance rather than devious intentions, the national rulers represented the landed interest. Their view was more generous than the gentry-dominated local authorities, but they shared the broad outlook of their interest. As great landowners, they felt threatened by popular disorders, although the danger was perhaps not as immediate as that which faced the lesser landowners. Indoctrination against middlemen and large farmers over at least twenty years conditioned them to welcome the discomfiture of these interests. The depth of popular response to the proclamation of 10 September startled them. But whatever their motives, the Ministry encouraged the local authorities, who used the middlemen and large farmers as scapegoats for the food crisis. Thus indirectly the central authorities played an important role in the diversion of the poor towards specific targets during the food riots of 1766.

4 The Provincial Rioters

Earlier chapters have shown that sudden fluctuations in the prices of provisions precipitated the hunger riots of 1766, but that social tensions had been mounting in rural England at least since the mid-century. This background of social instability coloured the perceptions of both the authorities and the labouring poor. Frightened at the prospect of a repetition of the events of 1756–7 when they found themselves faced with a hostile combination of the middle-class farmers and their labourers, the aristocracy and the gentry in 1766 seized the opportunity of diverting the lower orders away from themselves and towards the middlemen and the farmers. But the rural leaders were only able to achieve this manipulation, limited though it was, because of the resentments of the poor in the face of a sharp decline in their conditions after 1763, which followed five years when the living standards of the lower orders improved significantly.[1]

In addition to the social instability produced by agrarian developments after 1750, two major causes of the resentments among the labouring poor in rural England of the first decade of George III's reign, which ensured that the sudden fluctuations in food prices would produce a violent response, were the effects of the Seven Years' War and the trade disruptions of the later years of the 1760s. These two factors are evident from a study of the rioters of 1766.

I

The most significant element in any serious disturbance is the hard core of rioters who are willing to challenge initially the forces of order. This is the element which by its example convinces the more timid, who compose the majority in most mobs,

[1] Ashton, *Economic Fluctuations in England, 1700–1800*, p. 22.

that they may join with relative impunity.[2] In 1766 it was those familiar with military organisation and the military mind who were bold enough to face the hostility of the army and the magistrates. At this time there were three groups in particular who were used to arms, able to accept rudimentary discipline and knowledgeable in military tactics and who were among the most alienated of English lower-class society. These were the army, militia and navy veterans of the Seven Years' War.

In the surviving records of the 1760s few of the rioters are identified as veterans, and only by piecing together fragmentary evidence and drawing inferences from the tactics of the mobs, the dress of the rioters, the occupations of witnesses, and the apprehensions of the authorities can a picture of their important role be built up. Any study which related the home parishes of demobilised veterans to the riot areas would be most valuable, but the paucity of material precludes this.[3] Yet what evidence there is strongly suggests that ex-servicemen and militiamen played a significant role in giving shape and direction to the hunger riots of 1766.

Although hunger mobs, unlike riotous seamen, coalheavers and weavers, were rarely armed, several facets of their tactics speak of military experience. The restraint and honesty of the mobs in the early stages of the riots suggest a rudimentary organisation and leadership which would have been beyond the capacity of untrained rustics to provide. Local leaders enforced their authority upon those who failed to respond to the rallying call of the cow's horn or who disobeyed their orders.[4] In one incident, when rioters were accused of stealing from a farmhouse which they had just searched for food, for example, they submitted to a personal search and severely punished one of their number found with

---

[2] In the 1966 Cleveland riots the poor blacks and whites were encouraged to participate in looting and disorders when they saw the police did not intervene against the initial looters (Ken Southwood, 'Riot and Revolt: Sociological Theories of Political Violence', *Peace Research Reviews*, I, no. 3 [June 1967] 39).

[3] Such a study relating to the French Revolution was undertaken by Forest Macdonald, 'The Relation of the French Peasant Veteran of the American Revolution to the Fall of Feudalism in France', *Agricultural History*, xxv (October 1951) 151–61.

[4] *Treasury Solicitor's Papers*, T.S. 11/5956/Bx1128. *Marching Orders*, WO5–55, pp. 357–8.

some stolen spoons.[5] Inevitably as the disorders continued over several weeks in September and October of 1766, however, earlier restraints were dropped. In the subsequent trials rioters faced a variety of charges for personal assault, destruction of property and theft. Most intriguing were the alleged offences of a Wiltshire rioter who was accused of destroying a bolting mill and stealing four sack of meal, two Bibles, two Books of Common Prayer and a copy of *The Whole Duty of Man*.[6] Perhaps some took comfort in the reflection that men did not live by bread alone. Tactically mob organisers seem to have deliberately aimed at scattering the army into weak detachments by simultaneous risings and rapid movements of rioters. In Warwickshire for example on 6 October 1766 a mob of 1000 divided into gangs of 300 or 400 which simultaneously visited several neighbouring market towns.[7] Disciplined and concentrated volleys of stones which drove back the authorities in Norwich as the mob went about its work suggest an intelligent attempt to compensate for a lack of weapons.[8] At times the calculated actions of the rioters were reminiscent of a military campaign. Mob captains often chose targets ahead of time, although they also attacked spontaneously 'targets of opportunity'.[9] Occasionally mobs acted as armies of occupation. Leaders for example stationed their followers in private homes and inns during one extended period of disorders '*in the same manner as soldiers are billeted*'.[10]

Court records, too, suggest the involvement of veterans or militiamen by occasional reference to the wearing of oddments of military or naval uniforms by rioters. There are several such references in the Sessional Papers dealing with the Norwich riots of 1766, in the Norfolk Record Office, which are much more complete than the records of the agrarian disorders held by other county record offices. The authorities, for example, reportedly searched for Thomas Bear, a weaver who had served in the army

[5] John Pitt to Hardwicke, 29 September 1766, Add. MSS., 35607, fol. 290.
[6] *Treasury Solicitor's Papers*, T.S. 11/1116/5728. *Annual Register*, x (1767) 84.
[7] *Public Advertiser*, 20 October 1766. See also *Treasury Solicitor's Papers*, T.S. 11/5956/Bx1128 and T.S. 11/995/3707; and Barrington to the Earl of Suffolk, 1 October 1766, *Letter Book of Viscount Barrington*.
[8] Case, *Depositions and Case Papers* (1766).
[9] *Gazetteer and New Daily Advertiser*, 22 October 1766.
[10] *Treasury Solicitor's Papers*, T.S. 11/5956/Bx1128. My italics.

and had left his home parish in the regimentals of the eastern battalion of the Suffolk militia, and for Edward Shauter who had absconded from the 'Widow Cooper of Cosslang Bridge . . . with militia cloaths'. One witness of a riot at a Norwich malthouse claimed to have seen 'one of the rioters who was a very low man, and . . . not more than five feet high and had on a militia man's coat, go past the said house towards the above malthouse with a lighted lamp in his hand and that presently the said malthouse was set on fire and consumed'. The churchwardens of St John's, Timberhill, in Norwich, described another suspect who had absconded as one 'George Bambry, between eighteen and twenty, five feet five or six inches, with a militia coat and his own hair'. Even more significant was the fact that a leading witness in the Norwich Special Assizes was the Sergeant-Major of the Norfolk militia, for who would have been better able to identify rioting militiamen?[11] Yet the participation of militiamen and ex-servicemen was not peculiar to Norwich, as is evident from the real concern of the national authorities at the threat posed by riotous mobs led by such veterans.

Lord Barrington, an experienced and competent administrator who served as Secretary-at-War for almost the entire Seven Years' War and was destined to complete a further thirteen years in the same capacity before 1778, was preoccupied with this danger. Although a man neither by training nor temperament given to panic, Barrington decided at the height of the hunger riots in September 1766 that it was essential not to dissipate the army's strength by dispersing it into ever smaller detachments to meet each fresh outbreak: '. . . as a few soldiers commanded by a weak, ignorant subaltern might be defeated by a very large mob, full of men *lately used to arms in the army and militia. . . .*'[12] The War Minister believed that such an easily-gained victory might have encouraged scattered groups of rioters to unite into a general insurrection. He might have added that isolated detachments of troops heavily outnumbered by rioters of their own social class and led by former comrades were likely to throw in their lot with the rioters and a general insurrection might ensue, a nightmare possibility for all Ministers-at-War until the balance swung in

---

[11] *Depositions and Case Papers* (1766).
[12] *Shelburne Papers*, vol. 132, fol. 13–17. My italics.

favour of the forces of order, and Chartism declined in mid-Victorian England.[13]

That the War Office was highly sensitive to the possibility of disaffection among the troops is evident from the frequent orders for detachments to move from one billet to the next, presumably before the soldiers became over friendly with the local population and therefore unwilling to fire on civilian friends if disorders broke out.[14] Harsh punishments were the lot of riotous troops in Ireland and Wilkite London. The *St. James's Chronicle* reported the sentences of two soldiers who got drunk, joined the Southwark mob and roared 'Wilkes and Liberty for ever' : 'forty-five lashes each, three times, viz. every other day, and afterwards to drill for forty-five days. . . .'[15] Plainly not just Wilkites remembered 'Number 45, North Briton'. Nor was it coincidence that the authorities, who were well aware of the xenophobia of the London mob, chose Scottish troops to suppress the riot outside the King's Bench prison in 1768, or that Barrington collected broadsheets which tried to subvert the loyalty of the soldiers.[16] Not the least alarming aspect of the coalheavers' riots of 1768 was the suspicion that the Guards from the Tower were moonlighting as coalheavers with the connivance of their officers, who some believed were pocketing the soldiers' regular pay, and that at least one guardsman was an East End innkeeper deeply involved in the waterfront disputes.[17] The involvement of soldiers in this industrial dispute may partly explain its peculiarly violent character, and the bitterness between coalheavers and seamen. The traditional practice of English soldiers taking part-time employment caused economic rivalry between troops and seamen and led to riots in American at this time, and some of this rivalry may well have spilt over into England.[18]

If the ruling orders were concerned about the trustworthiness of the regular troops in moments of crisis, they were even more dubious about the reliability of the less-disciplined militia. The debates centring upon the new proposals for a militia early in

---

[13] Mather, *Public Order in the Age of the Chartists.*
[14] *Marching Orders, passim.*
[15] *St. James's Chronicle,* 12 May 1768.
[16] East Suffolk Record Office, Ipswich, *Barrington Papers.*
[17] See Part Two, Chapter 2, below.
[18] Lemisch, 'Jack Tar Versus John Bull'.

TO ALL

Gentlemen Soldiers and *Englishmen.*

By a Youth of Eighteen.

WHY sons of freedom, why suspect each other ?

Will half-fed soldiers kill his half-fed brother ?

Fear not my friends, the soldiers are as we,

English, freeborn, still longing to be free,

Tho' bold in arms, against our country's *foes,*

Britania's friends they mean not to oppose :

No vassals they, no fierce destructive band

Who'd stain with *kindred* blood their *native land,*

Abroad, tho' dauntless in the hostile field,

At *home* their souls to kind affections yield :

They rue with us the hapless fate we meet,

And will not *slay* because we want to *eat.*

" Know then this truth," An *English* soldier fights,

Not to *enslave*—but to defend our rights.

Printed for JOHN BULL.   1768.

1757 illustrate this questioning. One critic put his concerns this way : '. . . when the good people of England are thus armed and disciplined, they will be enabled to take away meat, corn, and malt, and other provisions, from the forestallers and ingrossers, butchers, millers and farmers, at a reasonable price, of which they themselves must always be the best and most impartial judges.'[19] His prophecy was given added appeal by the fact that riots over the militia muster sheets in the same year frequently did spill over into violent protests at the high price of food, when 'just prices' were enforced.[20]

The unsettling effect of militia service upon farm labourers and others of the rural population was always a cause of worry to the ruling orders in the English countryside. They believed with some justification that militiamen learned all the vices and few of the virtues of the regular soldiers, and returned to their native parishes with a taste for carousing and an aversion to hard work. When his regiment of militia was disbanding, Francis Russell, Marquis of Tavistock, wrote in December 1762 of finding work for his men to prevent their becoming 'bad subjects by being drove into idleness or that they should starve for want of employment'. He explained his actions in this way : 'The principal point I always laboured at of preserving their morals and not making them bad countrymen by disciplining them into good soldiers, has succeeded. . . . I never saw men to the last moment more orderly and well disciplined. . . . I own I am vain of their behaviour as soldiers but much more so of them as orderly, well-disposed men.'[21] Other landowners shared Russell's concern for the bad effects of service upon the poor labourers of their villages, but were rarely as concerned to assimilate them into rural life as he was.

Certainly military service had a significant effect upon rustics. Although under the 1757 Militia Act men did not serve more than a few miles away from their homes except in emergencies, many found themselves away from the home parish for the first time in their lives. With new horizons their expectations changed.

---

[19] *Gentleman's Magazine,* xxvii (1757) 132.
[20] Western, *The English Militia in the Eighteenth Century,* p. 300.
[21] Sir Lewis Namier and John Brooke, *The History of Parliament, the House of Commons, 1754–1790* (London : H.M. Stationery Office, 1964) iii 387.

They later found the transition back to civilian life difficult.

The changing attitude of the government to the use of the militia in times of riot in the first half of the eighteenth century reflected political and social realities. At the beginning of the century the militia rather than the regular army crushed popular risings.[22] This is partly attributable to the scarcity of troops, strong dislike of standing armies, and the non-social character of the riots. For dealing with political or religious riots, the militia, officered by the gentry and aristocracy, was at this time more trustworthy and certainly less provocative than the standing army. Suspicion of the regular army continued down to the French Revolution and more than one Whig in the 1760s bemoaned the growing reliance on 'that dreaded monster, the military' to suppress civil commotions. Despite this, by the midcentury the government relied more and more upon the army rather than the militia.[23] The collection of precedents for the use of military force by magistrates against mobs in the Earl of Shelburne's papers indicates that this procedure was relatively novel and that the authorities anticipated Parliamentary criticism for it.[24] To disarm such criticism the Secretary-at-War always took care to draft orders to military commanders to aid magistrates only 'upon their requisition'.[25]

This transition from the use of the militia to a reliance on the army to deal with disorders reflected the fact that rioting was more frequently the expression of social protest, rather than political faction, especially after 1750.

The reluctance of the authorities to use the militia to suppress the hunger riots of 1756–7 and particularly 1766–7 is readily understandable. The militia was composed of precisely the same elements as the hunger mobs. It was the industrious poor who rioted over high food prices in these years, and it was the industrious poor who filled the ranks of the militia. The Northamptonshire militia battalion in 1766, for example, consisted of labourers, servants, weavers, shoemakers, carpenters, shepherds, blacksmiths, woolcombers, gardeners, victuallers, braziers, masons, cordwainers,

[22] Max Beloff, *Public Order and Popular Disturbances, 1660–1714* (London: Frank Cass, 1963) *passim*.

[23] *Marching Orders*, WO5–54 and WO5–55.

[24] *Shelburne Papers*.

[25] *Marching Orders*, WO5–54, *passim*.

tailors, nailers and mat-makers.[26] There was some merit in news-paper criticisms of 'panic measures' to embody the militia because it meant 'arming the disaffected'.[27] Conversely the differing com-position of the army suggests why it was more reliable in times of social protest. In contrast to the militia which was more of a territorial force drawing on the country population, the army recruited primarily in manufacturing centres. If the army filled its units from the 'scum of the towns', it was a very different social grouping from the militia.[28] Such army recruits were prob-ably more malleable than the independent, lesser freeholders, small craftsmen and farm labourers who characteristically formed the backbone of the militia. Ferocious discipline and frequent moves between billets ensured that the army could be persuaded to fire on civilians in times of disorder.[29]

Another characteristic of the militia made it less suitable than the army for riot duty. The gentry-officers in the 1750s and 1760s were opposed to such use of their men. Seeking instructions on the attitude the militia should adopt to 'civil affrays' in 1760, Savile, an influential Yorkshire landowner, declared he would not wish to be set 'to play at soldiers'.[30] On another occasion West quoted Onslow who thought Barrington's proposal to use the militia to suppress popular disorders would 'disoblige' the officers and militiamen who did not consider such activity part of their job.[31] During the Parliamentary discussion of Lord Barrington's proposal to use the militia to maintain public order, Mathew Ridley expressed the attitude of many militia officers when he warned the government against the proposal in these terms : 'The military are to aid and assist the civil magistrate, not to war on the people. You are to employ the Militia so as not to disgust the Gentlemen employed in the Militia.'[32]

Probably another reason why the militia was not used against popular disorders was that the relations between the army and

[26] Western, p. 272.
[27] *Gazetteer and New Daily Advertiser*, 13 November 1766.
[28] Western, p. 272.
[29] *Marching Orders*, WO5–54 and WO5–55, *passim*.
[30] Sir George Savile to Rockingham, 31 July 1760, Rockingham MSS., R1–166.
[31] West to Newcastle, 17 May 1768, Add. MSS., 32990, fol. 98–9.
[32] British Museum, Egerton MSS., 215, fol. 58. See also Namier and Brooke, pp. 352–3.

the militia were rarely cordial. Whenever the county lieutenancy embodied militia battalions for training, a spate of orders left the War Office to clear regular troops from the mobilisation areas.[33] There was a natural antipathy between the two groups, related not only to their differing social composition but to such disparate causes as pay differentials, urban and rural rivalry, competition for feminine company and moonlighting. It may well be that another reason for the reluctance of the government to use the militia in 1766 was the fear of a bloody confrontation of the army and militia when social and economic discontents were dominant.

Distrust of amateur soldiers is also evident in the authorities' concern for the safety of weapons in times of disorders. Military equipment of the Middlesex and London militia battalions was lodged in the Tower on such occasions.[34] This practice of disarming militia and volunteer regiments in times of acute discontent was an offence to the pride of the volunteers especially for it indicated an official lack of confidence in amateur soldiers.[35]

Had the loyalty of the militia been beyond question, still their limited training and looser discipline made them less suitable for riot duty than the army. Facing hostile crowds of jeering and stone-throwing rioters required great restraint from outnumbered troops. The authorities found it necessary to build a stone wall to protect the troops in Gloucester market in October 1766,[36] and in the St George's Fields' riots of 1768 eighteen soldiers suffered severe bruises and cuts from the mob,[37] some of whom looked down the musket barrels and jeered at the troops because they had only loaded with powder and not ball.[38] The most valuable forces against rioters were the mounted dragoons of the regular army, who could break up large crowds without resort to firing, and could prevent their re-forming elsewhere. In the 1760s militiamen were all foot soldiers. The volunteer, yeoman cavalry

[33] *Marching Orders*, WO5–54, *passim*.
[34] Welbore Ellis to the King, May 1765, *Correspondence of King George the Third*, ed. Fortescue, pp. 108–9.
[35] Anthony Highmore, *The History of the Honourable Artillery Company of the City of London* (London, 1804) pp. 328 *et seq.*
[36] *Gazetteer and New Daily Advertiser*, 15 October 1766.
[37] 'Return of Men which were Hurt or Wounded in St. George's Fields by the Mob, during the Late Disturbances', *Barrington Papers*.
[38] *Treasury Solicitor's Papers*, T.S 11/920/3213.

Return of the Men which were Hurt or Wounded in St. George's Fields, by the Mob during the late Disturbances.

Corps.	Companies.	Mens Names.	How Wounded or Hurt?
2ᵈ Troop of Horse Grenadiers.		Richard Tatam	Cut on the Nose and Left Cheek very Deep.
		William Prout	Cut on the Right Cheek to the Bone.
2ᵈ Battalion 1ˢᵗ Regiment of Guards.	Earl Ligonier's	Corporal Figg	Some Bruises, — not Bad
	Lord Percival's	William Wadley	
	Lᵗ Colᵒ Miles's	Israel Morton	
	Lᵗ Colᵒ Thornton's	Joseph Bellinger	
	Lᵗ Colᵒ Hervey's	Cornelius Hanyon	
Coldstream	Colᵒ Evelyn's	Wᵐ Hood Drummer	Very much hurt in the Testicles
3ᵈ Regiment of Guards	Lᵗ Colᵒ Smith's	Serjeant Earl	Cut in the Face with a Stone
	Lᵗ Colᵒ Douglas's	Serjeant Byars	Lamed in the Legs & Arms with Stones.
	Lᵗ Colᵒ Twisleton's	Serjeant Dorrel	
	Lᵗ Colᵒ Faucitt's	Corporal Pratt	
	Lᵗ Colᵒ Tash's	Samuel Lowrey	Cut in the Hand.
	Colᵒ Hale's	Peter Barkley	Unmercifully bruised by the Mob
	Lᵗ Colᵒ Bridzleck's	Edmond Whiting	
	Lᵗ Colᵒ Jones's	John Mathews	Cut in the Legs.
	Lᵗ Colᵒ Smith's	Robert Leveret	
	Sir George Osborne's	Joseph Arundall	Cut in the Face.

which played such a notorious role in the 'Peterloo Massacre' was not formed until later in the eighteenth century.[39]

Although they did not use the militia against the hunger rioters, the authorities embodied most county battalions in the summer and autumn of 1766. The press reported the Berkshire battalion mobilising at St Alban's 'in the face of rioting', but usually the reason given for embodiment was to perform their annual training of twenty-eight days.[40] That the time was unusual and related to the emergency is apparent from the complaints in the press that militia mobilisations were interfering with the harvest.[41] There was criticism of 'panic measures' to embody the militia because it meant the arming of the 'disaffected'. Probably their mustering was as much for the purpose of placing under discipline men who would otherwise have reinforced the hard core of rioters, as for a last-ditch reserve for the regular army should a general insurrection develop. The tenor of the Parliamentary debate on Barrington's militia proposals in 1768 indicates that there was also some doubt about the legality of employing the militia against rioters.[42] It is unlikely however that these considerations of legality would have inhibited the authorities from using the militia in 1766, in view of the weight of precedent earlier in the century, and almost certainly in 1768 Lord Barrington was testing reaction to the prospect of stronger military suppression in face of the spreading disorders of that year without publicly committing the Ministry to what might have proved a highly unpopular course.[43] The frequency with which rioters proved to belong to the militia reveals that the policy of embodying militia units to control potential rioters was not wholly successful. These embodiments could not be main-

[39] See the Hammonds' comments on the untrustworthiness of the army and the militia in the 1790s and the use of volunteers (J. L. and Barbara Hammond, *The Town Labourer*, Paperback edition [London: Longmans, 1966] pp. 93–6).

[40] *Gazetteer and New Daily Advertiser*, 15 October 1766.

[41] Middlesex militia were called out for twenty-eight days' training, which was much resented at harvest time (ibid., 1 September 1766).

[42] Barrington proposed a measure which would have authorised the use of militia units to suppress civil commotions (Rockingham to Newcastle, 17 May 1768, Add. MSS., 32990, fol. 83–6).

[43] Brooke, *The Chatham Administration, 1766–1768*, p. 358. See also Newcastle to Pelham, 18 May 1768, Rockingham MSS., R1–1064; and Newcastle to Rockingham, 18 May 1768, Rockingham MSS., R1–1062.

tained in being for long, and once demobilised the militiaman could engage in social protests along with his fellows. By 1769 commanders were parading militia units for the purpose of identifying rioters. The following evidence was given against a silk-weaver accused of serious rioting : 'The prisoner Valline is a militia man and this boy singled him out from the whole corps when he was apprehended notwithstanding he was in a different dress.'[44]

In the eyes of the authorities, naval seamen were scarcely more reliable than the militia in times of riot in the ports. The reluctance of the admiralty to use naval seamen to suppress the merchant seamen's strike of 1768 was reflective of a concern on this account. The naval authorities on this occasion had to be content with blockading the Thames and other rivers, rather than actively suppressing the strikers.[45] In the eighteenth century the distinction between naval and merchant seamen was not always clear. During wars, many merchant seamen were quickly impressed by the navy. Thus when seamen struck for higher pay in 1768 they must have had considerable sympathy among the naval ratings assigned to prevent their escape to France. The involvement of ex-seamen in rioting of the period showed that the authorities' worries had some foundation in fact. Occasionally hunger mobs were reportedly led by men dressed as seamen, but more frequently seamen were involved in smugglers' disturbances, another form of social protest in the second half of the eighteenth century.

There were several ways in which the connection between smugglers' disturbances and hunger riots was apparent. First, the coincidence of serious hunger riots and the intensification of smuggling after the mid-century suggests they had a common basis in social protest.[46] While it is true that government tariff policies at various times encouraged smuggling of such valuable commodities as tobacco or brandy, the harsh conditions of life caused by scarcity and high prices when employment was scarce also drove the poor to find extra-legal sources of income. Probably the most likely to take quickly to smuggling were the seamen who found employment scarce after the Seven Years' War. Many of them had originally been in the coast-wise shipping trade

---

[44] *Treasury Solicitor's Papers*, T.S. 11/818/2696.
[45] See Part Two, Chapter 2, below.
[46] *Gentleman's Magazine*, xxvii (1757) 528.

before impressment by the navy. They readily applied their naval experience upon their return to civilian life. Second, areas where smuggling was rife were often regions where hunger riots occurred.[47] Frequently such hunger riots were more violent than those which took place elsewhere. For example, some of the bloodiest encounters between the troops and hunger rioters in 1766 were in the neighbourhood of Devizes, an important smuggling centre.[48] Third, undoubtedly frequent smuggling disorders created a climate of violence and a commonplace defiance of authority that were important elements in the background of some of the hunger riots of 1766. The role of these smuggling bands was roughly comparable to that of the poaching gangs in the East Anglian agrarian riots of 1816.[49] Fourth, the government had difficulty in enforcing its embargo on the export of grain after 26 September due to smugglers, and the army had to patrol the southern coasts to end this illegal traffic.[50]

Conscious of these threats posed by the veterans of the Seven Years' War, Lord Barrington urged Lord Shelburne, Secretary of State for the Southern Department and one of the Ministers responsible for internal peace in 1766, to concentrate the troops in each county under a 'good, prudent officer', even at the risk of giving up large tracts of territory to the rioters. Each Lord-Lieutenant should select one or two magistrates of 'spirit, activity, and discretion' to find among the rioters 'proper objects of punishment', issue warrants and arrest culprits with the aid of the military. Mobs interfering in this process should be 'chastised' by the troops, 'and the more roughly the better' for 'some bloody heads would be a real kindness and humanity'. The Minister-at-War advised the pacification of southern England county by county, starting with Wiltshire, then moving to Gloucestershire, and later dealing with the remaining counties in turn.[51] In the event, this plan was never implemented. Even the temporary abandonment to the mob of a large area of the countryside where many great estates were located would have alienated the nobility

[47] Canterbury, Exeter, King's Lynn, Norwich, Tiverton and along the coasts of Cornwall, Dorset, Suffolk and Sussex.

[48] *Gentleman's Magazine*, xxxv (1765) 94; xxvii (1757) 528.

[49] Peacock, *Bread or Blood*, pp. 50–1.

[50] On 26 September the export of grain was forbidden by the Privy Council (6 George III, cap. 5).

[51] *Shelburne Papers*, vol. 132, fol. 13–17.

and the gentry, whose support in the House was vital to the survival of the narrow-based Chatham Ministry.[52] But its very advocacy by one of the most experienced war ministers of the eighteenth century is a measure of the threat posed by the war veterans.

Why were such Englishmen among the most alienated of the population in 1766? Essentially their wartime experiences coloured their perception of social conditions which had been deteriorating for many of the poor since at least the mid-century. Where before they had been willing to accept passively their lot, their military experience gave them a new perspective.

Many returned to civilian life with a sense of grievance which was aggravated by the conditions they found awaiting them. Few had enlisted during the war other than from necessity. Some had been impressed by the navy or army; others joined the armed forces as an alternative to starvation in the early years of the war when prices were very high, or to hanging for their participation in the 1756–7 food riots.[53] In some instances the authorities had broken promises made before enlistment in the militia for home service only. As noted above, the forcible embarkation of the Somerset and Dorset militiamen for service in America in 1756 had stimulated widespread opposition to a new Militia Act which erupted in the serious riots which spread across the Midlands and the north of England.[54] Only the better harvests in the south, generally lower prices of provisions and limitation of grain exports account for the fact that these riots did not spread southwards in 1756–7. A dearth of oats due to military requirements also affected the northerners, whose diet included 'crowdie' made from oats.[55] After the war the veterans found conditions little improved. Upon their demobilisation, the government presented the servicemen with licenses to beg their way home to their native parishes in lieu of a gratuity. There, frequently, they found their old jobs taken by others. During the war, industries had often increased production with a reduced labour force by raising pro-

[52] Brooke, p. 4 *et seq.*

[53] *Gentleman's Magazine*, xxvi (1756) 447 and *passim*; xxvii (1757) *passim*.

[54] Western, p. 122.

[55] Fay, *The Corn Laws and Social England*, p. 4; and *Parliamentary History of England*, p. 461.

ductivity. In Norwich, for example, despite the fact that 4000 men served in the war, production of cloth actually increased. Many war veterans, then, faced several years of difficult adjustment after peace was signed in 1763.[56]

Initially the government had facilitated the rehabilitation of mariners and soldiers by suspending the monopolies enjoyed by various guilds and livery companies, and permitting veterans of the Seven Years' War to set up in trade anywhere in Britain or Ireland under an act of 22 George II. In an effort to make the new Militia Act more palatable to those liable to service, Parliament extended this privilege in 1757 to married men who had served in the militia overseas or during an invasion or general insurrection. Another act of 1763 enabled ex-seamen to work on the river Thames and in other trades of the city.[57] As a result of these different acts, veterans crowded into some trades which were facing declining business. The watermen for example had already lost business to new bridges, and in May 1768 they assembled outside the Mansion House to seek the Lord Mayor's support for a petition asking Parliament for relief.[58] The position of ex-servicemen further deteriorated in 1766 owing to a legal judgement given in favour of the Farriers' Company which re-established a monopolistic principle favourable to the trades rather than to unemployed veterans.[59] As a consequence of this, demobilised soldiers and seamen lost the right to establish themselves in trades without the customary apprenticeship.

At a time when returned servicemen were finding employment increasingly difficult to obtain, prices of provisions were rising steeply. Since 1764 the prices of meat, dairy produce and grain caused the government concern, and in the following two years several committees of Parliament met to examine the causes of high prices.[60] Thus many who had entered the armed forces at a time of scarcity and high prices faced the very same problems they had sought to escape less than a decade earlier. In 1766 they were less inclined to accept their fate stoically.

[56] Arthur Young, *The Farmer's Tour through the East of England*, 4 vols (London: W. Straham, 1771) II 77–8.
[57] See Part Two below.
[58] *Gentleman's Magazine*, xxxviii (1768) 242.
[59] Henry Humpherus, *History of the Origin and Progress of the Company of Watermen and Lightermen of the River Thames* (London, 1887) II 262.
[60] In the years 1764, 1765 and 1766.

During their service careers the expectations of the poor had changed, not least in regard to their diet. While in the army or navy, they often endured an inferior diet due to the incompetency, parsimony or peculations of contractors and victualling officers, and doubtless few required the encouragement they certainly received from gentry, peers, industrialists and the Ministry to settle old scores with middlemen. Yet even those forced on occasion to eat inferior pork produced from hogs fed on fermenting distillery mash had frequently enjoyed a higher standard of living than they had known in civilian life.[61] Many ate wheaten bread for the first time in their lives. General Ligonier, as Chief of Staff at the War Office, had in 1743 introduced wheaten bread into the regular diet of the army for the first time. Writing subsequently of this change, he noted,

With regard to the article of bread I [attended] their Lordships in 1743, when my sentiments differed from General Honeywood's on that head. That it was true the troops had rye bread in the time of Queen Anne but that I believe more men were lost by this kind of bread than by the sword of the enemy, and therefore I recommended to their lordships, that they would be pleas'd to feed the troops with bread made of wheat only, which was complied with.[62]

Ligonier's reasons contrasted sharply with the arguments of his contemporaries in favour of the higher nutritional value of coarse grains compared with wheat.

That the step from rye to wheaten bread was a most significant one in the eyes of the poor of any European country, and one not easily retraced, is quite apparent. By the 1760s wheaten bread was the staple of most of the poor in southern England, although in the West Country coarse grains were eaten more frequently.[63]

[61] *Gentleman's Magazine*, xxvi (1756) 625.
[62] Ligonier to Samuel Martin Esquire, Secretary, The Lords of the Treasury, London, 15 July 1758, William L. Clement Library, Ann Arbor, Michigan, Ligonier MSS., *Letter Book* (1759–60).
[63] Fay, *The Corn Laws*, pp. 4–5, discusses the eating habits of the British using the evidence of Charles Smith, the *Parliamentary History of England* and of Eden. He concludes: (1) rye and barley bread rivalled wheaten in the Midlands, and was eaten in Wales almost solely; (2) oats were eaten in the north Midlands, the north of England, and Scotland; (3) potatoes in Lancashire; (4) southern counties ate predominantly wheaten bread. He calculated $3\frac{3}{4}$ of 6 million population in 1766 ate wheaten bread. He accepts Charles Smith's figures in *Three Tracts on the Corn Trade and the Corn Laws* (1766).

In the north potatoes together with oats and barley figured large in the diet of the poor.[64] Rice too was popular, and Charles Townsend supported a petition of eminent grocers to ban the export of rice, due to the high prices of rice in Holland, which threatened to drain the country and distress the poor.

This conservative attitude of the poor towards their diet is evident from a study of the prices of grains published monthly in the *Gentleman's Magazine* during 1766. In this year of food crisis the prices of coarser grains remained remarkably steady, while the prices of wheat fluctuated violently. This price pattern was partly due to the fact that dealers and farmers did not offer barley and rye for sale in the open market normally, although the press quoted the prices of such grains as if they could be bought by the public at those prices.[65] Maltsters forestalled the market at Gloucester and Tewkesbury to supply the brewers and distillers of Bristol, for example.[66] Yet had there been a great demand from the poor for cheaper grains, dealers and farmers would have offered them for sale to the bakers and to the public. Prices then would have reflected this new demand. One newspaper did report a twenty per cent increase in barley prices within a fortnight in the autumn of 1766,[67] but the *Gentleman's Magazine*'s more complete tables of prices indicate no general price increase for barley or other coarse grains. This was despite the fact that many of the upper classes tried hard to encourage the poor to eat coarser bread. Lord Tavistock reported to his father, the Duke of Bedford, that he was mixing rye with wheat to sell at low prices for the poor to make 'nutritious' bread.[68] Newspapers too provided instructions on how to prepare wheat substitutes. But the poor generally seem to have been slow to accept barley, oats, beans or peas, or even to eat household bread which was made from wheat and which was believed more nourishing and longer-keeping than white bread.[69] This stubborn retention of newly-acquired eating

[64] *Considerations on the Exportation of Corn*, p. 63.

[65] Jacob Rowe to Lord Abercorn, 9 March 1765, *Committee on High Prices of Provisions*.

[66] A bailiff to Lord Abercorn, 8 March 1765, ibid.

[67] *Gazetteer and New Daily Advertiser*, 18 October 1766.

[68] *Correspondence of John, Fourth Duke of Bedford* (London: Longman, 1846) III 346–8.

[69] A modern example of conservative tastes in food was India where peasants rioted over the dearth of rice although American wheat was available; the former was culturally required (Southwood, 'Riot and Revolt', p. 42).

habits caused much comment from the wealthier citizenry. Jonas Hanway noted: 'It is not always enough that the provisions be good; it must also be of a particular kind.'[70] Another writer sensitive to the importance of the expectations of people as a factor behind unrest observed: 'We may calmly discuss matters over a bottle of claret after a plentiful dinner, and say that the poor in Ireland live on potatoes, and in France, and other countries upon turnips or cabbage. We must take these things as we find them, our poor are not accustomed to live in that means, nor will they give up bread. . . .'[71] The people of southern England, where riots were extensive in the summer and autumn of 1766, were said to have lost their 'rye-teeth', and did not wish to regain their taste for coarser bread.[72] They preferred to take action against the middlemen and large farmers, who they were willing to believe had brought about an artificial scarcity in time of plenty by their speculations.

The results of the Seven Years' War, then, were apparent in the involvement of veterans in the riots of the 1760s. Primarily in its effect upon the expectations of these men, the war had serious repercussions upon English society. It also had a serious effect upon the economy. By creating an inflated demand for certain goods such as textiles and hardware products, the war caused a rapid expansion of the labour force in certain industries which had to cut back upon the resumption of peace. At the same time the war exacerbated the realignment of trade which was occurring after the mid-century. These effects are evident from a further study of the agrarian rioters of 1766 and their specific grievances.

II

Unemployment and underemployment were the visible effects of declining trade after 1764.[73] With the advent of peace came

---

[70] *Gazetteer and New Daily Advertiser*, 19 November 1766; and Jonas Hanway, *Letters on the Importance of the Rising Generation of the Labouring Part of our Fellow-Subjects* (London, 1767) I xxxii 194.

[71] *Gazetteer and New Daily Advertiser*, 1 November 1766.

[72] Fay, *Corn Laws*, p. 4 et seq.

[73] Elizabeth Boody Schumpeter, *English Overseas Trade Statistics, 1697–1808* (Oxford: Clarendon Press, 1960) Table II, p. 15. See also Ralph Davis, 'English Foreign Trade, 1700–1774', *Economic History Review*, 2nd ser., xv (1962) 285–303.

reduced government contracts for manufactures and victualling stores. Industries such as hardware, textiles, coal-mining and ship-building, which had expanded rapidly under the stimulus of war, had now to contract to peacetime size. Contemporary commentators noted the distress in two of these industries in particular.

Around the Midland centres of Birmingham, Walsall, Wolverhampton and the south Yorkshire city of Sheffield, distress among the hardware workers was very great.[74] Related industries such as coal-mining felt the effects of reduced working in the iron industry too. Attempts by coalowners to cut costs by returning to the lower wages current before the war increased unrest in the coal fields of the north-east, already stirred by the indenture system which was operated by the employers to ensure an adequate labour supply.[75] It is not coincidental that Bristol colliers were active in regulating food markets in 1766, and that the authorities greatly feared their disciplined challenges to public order.[76]

The post-war decline of the cloth trade had an even greater effect on the conditions of the poor because of the widespread nature of cloth production. That part of the industry which was organised on a 'putting-out' system in good times provided a supplementary source of income to the rural poor which put them above subsistence level. Competition from the new worsted industry of Yorkshire for the shrinking post-war market in cloth forced an acceleration of the reorganisation of the cloth industry in the old cloth counties of southern England,[77] with its attendant misery for individual cloth workers.[78] Even the new worsted centres in Yorkshire felt the impact of trade depression by December 1765. Observers estimated that most of the 500,000 cloth workers in Leeds, Wakefield, Bradford, Keighley, Halifax, Huddersfield, Rochdale, Morley, Burstall, Batley, Pudsey, Dewsbury, Ossett, Kirkheaton and Almonsbury were unemployed or on half-time by this date.[79] But the recession of the cloth trade was most severe

---

[74] *Anuual Register*, IX (1766) 61.

[75] *Gentleman's Magazine*, XXXV (1765) 430, 488; and *State Papers*, SP 37/4, fol. 1.

[76] 'Letter from Bristol', *Public Advertiser*, 2 October 1766.

[77] *Victoria County History, Gloucestershire*, II 160 and *passim*. *Victoria County History, Wiltshire*, IV 62 *et seq.* Heaton, *The Yorkshire Woollen and Worsted Industries*.

[78] Lipson, *History of the Woollen and Worsted Industries*.

[79] *Gentleman's Magazine*, XXXV (1765) 567.

in the southern counties, which helps to explain why the hunger riots were most serious in the old cloth counties of Wiltshire, Gloucestershire and Norfolk.

Cloth workers were frequently involved in the riots of 1766 in the West Country and East Anglia, as is evident from court records and contemporary writers. One shrewd commentator noted : 'The first riots in Gloucestershire were occasioned by the discontents of the lower sort of labourer in the clothing trade, who, finding the work scarce and provisions dear, grumbled till a hint given in the public newspaper was recommended to them by their employers of regulating the markets.'[80] Another correspondent blamed the desperate state of the poor on a minority of troublemakers who rioted without justification and hampered the movement of grain into the essentially pastoral western counties by regulating markets and frightening away the farmers from neighbouring counties. He observed : 'The rioters, for the most part, were sturdy fellows chiefly weavers, scribbers, and shearers who could have earned 9–30/- if they worked. These forced down the prices and took to terrorising farmers and ale-house keepers for food, beer, and money.'[81] Elizabeth Gilboy has suggested that the cloth workers took the opportunity to repay old scores with clothiers at this time, but they attacked bolting mills where flour was sifted rather than the property of their employers. Discounting drunkenness of manufacturers as an explanation of the disorders of 1766, Dr Gilboy suggests the workers remembered the wage struggles of 1727 and 1756, and wished to make things as unpleasant as possible for their masters. While this does not account for the cloth workers' attacks on the property of middlemen and farmers rather than that of wealthy clothiers, it does suggest a reason why their employers wished to encourage them to force down prices.[82] James Bryant, reporting to Lords Lovell and Holland regarding the questionnaire sent by the House of

[80] John Pitt to Hardwicke, 21 December 1766, Add. MSS., 35607, fol. 341. By the summer of 1768 there was some improvement in the West Country clothing industry (*Public Advertiser*, 17 June 1768).

[81] Only a minority of the 30,000 or 40,000 cloth manufacturers within thirty miles of Minchinhampton and the vicinity of Stroud were said to be troublemakers. Most were poor men, earning eight or nine shillings per week as 'sober, industrious workers' (*Gazetteer and New Daily Advertiser*, 22 October 1766).

[82] Gilboy, *Wages in Eighteenth Century England*.

Lords' *Committee on High Prices of Provisions*, noted the practice of local millers, 'who of late years set up bunting mills to go about farms buying wheat in the rick which they grind [to] make into flour—thus much wheat [is] prevented from the market and resulting dear prices'. Suspicion that millers were adulterating the flour further increased the antipathy of the poor. When stocks of alum, whiting and other adulterants were found in a bolting mill, the mobs became very destructive.[83] Certainly with the wage disputes of the 1750s less than a decade earlier, it is hardly surprising that clothiers diverted the attention of their workers towards middlemen and large-scale farmers, the apparent authors of an artificial shortage of food, as John Pitt told Lord Hardwicke.[84] There can be little doubt that cloth workers suffered reduced employment due to a general realignment of trade in the immediate post-war period.[85] Distress was widespread across southern England when agrarian hunger riots broke out in 1766.

But the distress of the industrious poor of English manufacturing centres, as well as of the farm labourers, was due to trade disruptions which had little to do with any post-war recession. One major cause of such disruptions was the American colonists' response first to the Stamp Act and later to the Townshend Duties.[86] Non-importation agreements in America came into operation in 1765 and reduced loadings for New England, New York, Pennsylvania, Maryland, Virginia, the Carolinas and

---

[83] James Bryant to Lords Lovell and Holland, *Committee on High Prices of Provisions*.

[84] After the weavers' riots of 1756, justices of the peace in Quarter Session had again fixed wages, but clothiers circumvented the regulations, pleading competition from the French was due to the export of English and Irish wool (*Victoria County History, Gloucestershire*, II 160).

[85] Until the last decade of the eighteenth century English cloth was sought abroad for its cheapness rather than its quality, e.g., there was a decline in export of English cloth to Portugal after the Seven Years' War due to a failure to raise the quality for the price asked. After 1770 there was a marked shift from coarser to finer cloth (*Documents Illustrating the Wiltshire Textile Trades in the Eighteenth Century*, ed. Julia de L. Mann [Devizes: Wiltshire Archaeological and National History Society, Records Branch, 1964] XIX). A letter from Wiltshire of 20 September referred to the distress of out-of-work cloth workers (*Gazetteer and New Daily Advertiser*, 26 September 1766). The *Public Advertiser*, 19 September 1766, reported farmers dropping wheat to 6/- a bushel, 'especially in the clothing towns'. A letter from Gloucester reported clothing workers involved in the riots (*Gazetteer and New Daily Advertiser*, 22 October 1766).

[86] 'Letter from Bristol', *Public Advertiser*, 2 October 1766.

Georgia were the evident effects of the economic war.[87] In addition to the outright decline of trade with America, English merchants suffered from the non-payment of outstanding debts, which forced them to reduce orders to industrialists. Stocks piled up, and manufacturers laid off large numbers of workers, at a time when the prices of provisions were high. As the *Gentleman's Magazine* observed, 'Besides, as the poor are so numerous it will be very difficult to find employment for them, especially in manufactures for foreign exportation; as the prices of all sorts of provisions are now greatly increased.'[88] Annual figures of British exports to the Thirteen Colonies provide a measure of American hostility towards the Stamp Act and the Townshend Duties, and suggest the difficulties of the English workers: 1764, 2.49 (£ million); 1765, 1.94; 1766, 1.80; 1767, 1.90; 1768, 2.16; 1769, 1.34.[89] The years 1765, 1766 and 1769 were years when non-importation orders operated.

Some of the trade losses with America were made up by greater exports to Africa, the East Indies, the West Indies and northern Europe.[90] But certain industries did not benefit from this shift in trade. Certainly woollen textiles had little potential for expansion in tropical and semi-tropical countries. The overall tendency of trade in the second half of the 1760s was towards depression. The highest value of exports, including re-exports, was in 1764, the first year of peace when trade was free of restrictions. Exports as well as re-exports declined significantly in the next three years: 1765, 14,573 (thousands of pounds); 1766, 14,082; 1767, 13,867. In 1768 there was some recovery with 15,120, followed in the next year by a slump to 13,438 as reaction to the Townshend Duties took effect.[91] Exports of English produce and manufactures (without, that is, re-exports) do not even show a recovery in 1768: 1764, 11,536 (£'000); 1765, 10,122; 1766, 9900; 1767, 9492; 1768, 9695; 1769, 8984.[92]

[87] Ashton, p. 164.
[88] *Gentleman's Magazine*, xxxv (1765) 84–5.
[89] Ashton.
[90] Schumpeter, Table V, p. 17.
[91] Ibid., Table I, p. 15.
[92] Export figures of English produce and manufactures are more indicative of employment conditions within the country. Re-exports only marginally affect the home labour force and rather reflect conditions in the producing country. Sugar refining would have some slight effect on employment in the port economies of London and the western outports (ibid., Table II). See Ashton.

Trade figures do not show conclusively whether there was an internal depression or not in the 1760s, but they do reveal a significant reduction in exports and suggest reasons for the widespread unemployment in certain regions of the country reported in the press. One may reasonably conclude with Professor T. S. Ashton that the years 1765–70 were years of depression, and that they were the beginning of a twenty-year period when overseas trade declined greatly.[93] Ashton has explained the two decades of depression in terms of the economic conflict with America, the economic malaise of Germany, disorders in India, war with France and Spain and a reduction in shipments on the government account.[94] The American difficulties of the 1760s were part of a wider realignment of trade in the eighteenth century, the effect of which was to distress considerable segments of the industrial population and encourage them to riot in times of severely fluctuating prices such as 1766.[95]

But the agricultural labourer, too, felt the adverse effects of disruptions in trade after the Seven Years' War. Not only did the cottage industries like cloth-weaving decline with the reduction in overseas trade, but they declined also because English industrial workers had less money to buy rural goods. As already noted, natural shortages of food benefited the farmer rather than the labourer. After 1763 there were several seasons of epidemics among farm animals and a series of poor harvests. These natural disasters adversely affected the labourers, who had less work at a time of high food prices. Smaller farmers too suffered because natural disasters might be local and because poor rates were high. In some instances small farmers dropped into the labouring class as a result of these problems.

In the post-war years farm wages failed to keep pace with prices, and in the West Country they actually dropped. The generally depressed state of farming in the west was reflected in the practice of paying labour in kind rather than money. Farmers

    [93] Schumpeter, p. 4 (Introduction by T. S. Ashton).
    [94] Ibid.
    [95] David Macpherson, *Annals of Commerce, Manufactures, Fisheries and Navigation . . . Containing the Commercial Transactions of the British Empire and Other Countries, from the Earliest Accounts to . . . January 1801; and Comprehending the Most Valuable Part of . . . Mr. Anderson's History of Commerce, Viz. from the Year 1492 to the End of the Reign of George II, etc*, 4 vols (London, 1805) III 442–3.

based the quantities of produce paid as wages on the quantities appropriate to the early years of the century when food prices were low. When prices of provisions rose steeply in the 1760s, farmers reportedly paid labourers in corn 'which at the late enormous price . . . was not sufficient to support their starving families three days in the week'.[96]

Consequently farm labourers appeared frequently within the ranks of the hunger rioters of 1766, despite their traditional deference to privilege and the natural difficulty of organising a group of workers who were so geographically scattered.

Some contemporaries blamed the industrial workers rather than the farm workers for disorders. Arthur Young, hardly an impartial observer, claimed that the manufacturing classes, despite their high wages of nine to eleven shillings per week, were more turbulent than farm labourers who earned only five and six pence. 'Who would not conclude the same manufacturers the most unreasonable, tumultuous, and ungovernable people possible seeing others could be content who earned but half the money. [he asked] . . . But let not these lawless plunderers, who are universally the very skum, and riff-raff of their neighbourhood, have the least effect upon your opinions. The more such fellows earn, the more succeeding time and money they have for the ale-house and disorderly meetings. and of course, more in their power to do mischief.'[97] Few of Young's critics denied that industrial workers were less tractable than farm workers, but they rightly noted that wage rates themselves were not valid indices of living standards. In fact, the manufacturers were frequently unemployed from four to eight months of the year in the 1760s and their annual income was below that of farm labourers,[98] quite apart from perquisites of cheap food which most agricultural workers enjoyed.[99] Despite their relatively better living standards, the farm labourers joined the industrial workers in their social protests of 1766.

---

[96] *Gazetteer and New Daily Advertiser*, 29 September 1766.
[97] Arthur Young, *A Six Weeks' Tour through the Southern Counties of England and Wales* (London, 1768) pp. 330–3.
[98] See 'No Rioter' reply to Young's *A Six Weeks' Tour* (*Lloyd's Evening Post*, 25–27 May 1768).
[99] Gilboy, *Eighteenth Century England*, p. 80.

The fact that both industrial workers and farm labourers rioted in 1766 and thereby expressed an accumulation of grievances which had built up since the half-century, and had reached serious proportions due to the effect of the war and trade disruptions of the 1760s did not mean that other important groups were not involved despite their absence from the records.

Law enforcement in the eighteenth century was frequently haphazard and always had a distinct class character. There is much evidence to show that the courts gave heavier sentences to lower-class hunger and industrial rioters than to middle-class political rioters.[100] Farmers who rioted in 1756–7 and 1815 received milder treatment than the labourers they incited to riot in subsequent years by their actions.[101] It is reasonable to suppose that those arrested for their part in the 1766 disorders suffered because of their high 'visibility factor'. Just as today North American Indians or Negroes stand a greater chance of arrest for minor traffic offences than do middle-class Caucasian drivers, the poor of the eighteenth century in the homespun smocks or remnants of military clothing stood out clearly from their social superiors and were more prone to arrest. The inability of magistrates to arrest culprits during the course of most riots in 1766 encouraged this selective 'justice'.

The delay between the committal of the offence of rioting and the arrest of participants is significant in considering the composition of mobs. Magistrates were often unable to arrest rioters for several days or even weeks after their alleged offences. Without military assistance the authorities were unable to make summary arrests. As the army gradually restored calm to the rural areas in October and November 1766, the magistrates concentrated their efforts on hunting down the ringleaders of the mobs. Finding witnesses, taking depositions, and finally tracing offenders were all time-consuming and the justices frequently enlisted the aid of the churchwardens and other parish officials. They asked for information about the names of known rioters, villagers who were absent from their home parishes during the riots or people who had subsequently absconded.[102] Consequently those indicted

---

[100] Rudé, *Wilkes and Liberty*, Appendices III and XI.

[101] Peacock (p. 127) notes similar preferential treatment for farmers rioting compared with labourers rioting.

[102] *Depositions and Case Papers* (1766).

for rioting were almost invariably local men.[103] Prisoners tried before the four special commissions sitting in the chief disaffected counties of Gloucestershire, Berkshire, Wiltshire and Norfolk in December 1766 were not necessarily a representative cross-section of the rioters.

One important group almost certainly involved, at least in the early riots, of whom no record remains, was the seasonal harvest workers from London and other large urban centres.[104] More aware of the conspicuous consumption of the increasingly wealthy, urban 'middling sort', as well as of the aristocracy, they entertained higher expectations than many of the rural poor,[105] while at the same time they lacked the latter's traditional deference to privilege. Doubtless the concentration of population into the Tower Hamlets and the poorer liberties of London placed such an intolerable strain upon the parish system that magistrates found it impractical to trace harvest workers who rioted through parish officers in the same way as they were able to do in rural and smaller urban centres.

Probably within the ranks of seasonal workers were domestic servants between masters. These were among the most alienated groups in eighteenth-century England. Close proximity to the affluent had bred contempt and whetted their appetites for expensive luxuries of life which they could never afford. Originally brought to London from the countryside, they had quickly been spoilt by the alien society on whose fringes they now hovered, at least in the view of their masters. Frequently discharged as incorrigible by their employers and given no references, they resorted to deception to gain new employment. Those who did not support themselves by crime passed themselves off as newly-arrived, unsophisticated rustics in the city for the first time in search of employment.[106] The large number of unemployed servants between jobs was the cause of great alarm to the metropolitan magistrates. Riots of armed footmen and others over attempts to

---

[103] Cf. Rudé, *The Crowd in History*, for observations on the 'faces in the crowd'.

[104] Ashton, *An Economic History of England*, p. 32.

[105] Hanway claimed domestic servants ate meat three times a day and deprived the poor of all but three or four ounces (Hanway, *Letters*, xxxii, pp. 194 *et seq.*).

[106] J. Jean Hecht, *The Domestic Servant Class in Eighteenth Century England* (London: Routledge & Paul, 1956) *passim*.

discontinue their 'vails' in the 1760s indicate the militancy of this servant class.[107] Servants were identified among the hunger riots of 1766, but probably the more rootless members of this interest played an important role in both urban and rural riots out of proportion to the numbers who were identified. In urban riots particularly, servants were able to escape arrest because they dressed as gentlemen, and doubtless were able to bluff their way through in circumstances where a more readily identifiable member of the poorer sort would have been arrested.

Another group, reference to which is not found in the records of 1766, was the Irish harvest workers. Seasonal labourers from Ireland were not yet appearing in the English countryside in the large numbers characteristic of the early nineteenth century, but by the 1760s their numbers were growing. Energetic suppression of the terrorist gangs of Whiteboys by the Irish government had caused many to flee to England after 1765.[108] Most headed for the crowded parishes of east London where the authorities found it impossible to enforce the laws of settlement. As will be seen later, some established an Irish 'mafia' which dominated the waterfront of the Thames in May 1768. Yet identification with the land was always a strong driving force for Irishmen, and many must have found work in the harvest fields. It is surprising that little or no mention is made of Irishmen in the hunger riots of 1766. They would have certainly stood out from others of the poor because of their speech and dress. Certainly the authorities would have been ready to accept any suggestion of a conspiracy behind these dangerous disorders had there been any shred of evidence to support such a theory. Perhaps, however, the diversion of the rioters against middlemen and large farmers had frightened the authorities by its success. The subsequent threat to the social structure shocked them into the realisation that the precipitating cause of the riots was hunger. In such circumstances the search for another scapegoat was a waste of time. It is reasonable to suppose that Irishmen contributed their own resentments to the pressure for social change in England, and that many disaffected Irishmen were in the ranks of the rioters in 1766.

---

[107] M. Dorothy George, 'The Early History of Registry Offices', *Economic History*, Supplement of *Economic Journal*, I, no. 4 (1926–9) 229–48.
[108] *Correspondence of King George the Third*, ed. Fortescue, p. 310.

III

The composition of riotous mobs in 1766, their organisation, their tactics and their targets all expressed the underlying social tensions of England which became acute in the early 1750s. The impact of the Seven Years' War and the trade disruptions of the 1760s exacerbated these pressures in rural society. The fluctuating prices and apprehensions of outright famine touched off the violent protests, which subsequent government policies and magistrates' sanction encouraged almost to the point of general insurrection. Denied effective government control of wages and food prices, and often faced with more restrictive attitudes towards the granting of poor relief,[109] the poor allowed themselves to be diverted towards the middlemen and great farmers by a landed interest determined at all costs to avoid the social isolation it had experienced in 1756–7 when mobs of farmers and labourers united against it.

[109] Dorothy Marshall, *The English Poor in the Eighteenth Century: A Study in Social and Administrative History* (New York: Augustus M. Kelley, 1926) p. 14.

PART TWO

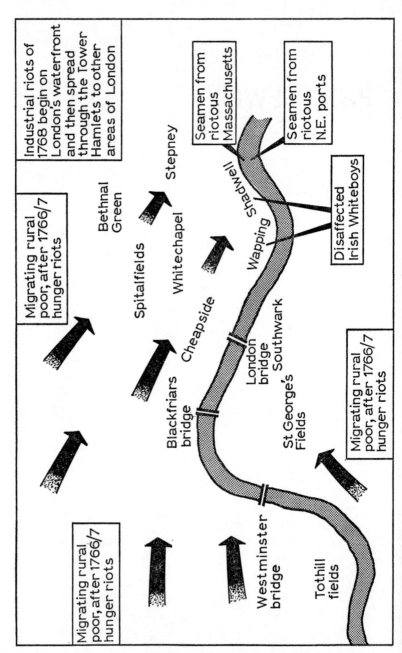

Industrial riots of 1768 begin on London's waterfront and then spread through the Tower Hamlets to other areas of London

Migrating rural poor, after 1766/7 hunger riots

Seamen from riotous Massachusetts

Seamen from riotous N.E. ports

Disaffected Irish Whiteboys

Bethnal Green

Stepney

Spitalfields

Whitechapel

Shadwell

Cheapside

Wapping

London bridge

Blackfriars bridge

St George's Southwark

Migrating rural poor, after 1766/7 hunger riots

St George's Fields

Westminster bridge

Tothill fields

Migrating rural poor, after 1766/7 hunger riots

London during the industrial riots of 1768

1 Introduction

Despite the serious hunger riots which extended through southern England in 1766, no comparable disturbances over high prices of food occurred in the London area in that year. Prior to 1768, the only serious metropolitan disorders in the 1760s were those related to the causes of John Wilkes in 1763, and to the discontents of the Spitalfields silk-weavers with government trade policies, which culminated in the seige of the Duke of Bedford's residence in 1765. Other disturbances among such groups as the seamen, shoemakers and footmen were of much more limited duration.

There were several reasons why the London populace refrained from violent protest when hunger mobs disrupted much of the surrounding countryside. Metropolitan magistrates were usually quick to safeguard supplies of food to the capital. Because of the complexity of feeding such large numbers of people and the obviously essential role of middlemen in this process, antipathy towards this interest was not as intense in London as it was in the provinces, where marketing systems were simpler. Although in times of emergency the London magistrates kept a close watch for speculative actions of middlemen which unfairly raised the price of 'necessaries' in times of scarcity, the old statutes against forestalling, engrossing and regrating by the 1760s were never as rigorously enforced in the more important urban centres as they were in the country at large. Hence neither the government's proclamation of the old anti-middlemen statutes in September 1766, nor the coincidence of high food prices and low wages in 1768 stimulated attacks on the middlemen of the Metropolis. Neither did grain movements in London and Westminster attract the same attention as they did in the provinces. Food was always arriving in the London area to feed the populace, as well as for shipment overseas. The importance of grain movements through

155

the countryside in precipitating riots in 1766 is evident from the correlation of disorders and changes in government policies on the export of grain. Although most grain shipped abroad in 1766 left through the port of London, its movement through the countryside to the capital, rather than its loading aboard vessels in the Thames, attracted the hostility of the poor. Yet another significant reason for the calm of the London populace in 1766 was the fact that the prices of food, especially grain, did not fluctuate wildly, as they did in the provinces. At the same time that prices remained relatively stable in the London food markets, the economic recession was not as severe in the capital's trades as it was in the clothing centres of the south, or the hardware industries of the Midlands. Indeed the repeal of the Stamp Act in March 1766 had a more immediate effect upon the great commercial centre of London than did the government's embargo on grain exports. With the exception of the silk-weavers and watermen, most London tradesmen were not yet suffering from large-scale unemployment. The industrious poor of London, therefore, did not face high food prices at a time of low employment, as did their provincial counterparts.

By early 1768, the conditions of many of the metropolitan poor had changed distinctly for the worse. As a result of continued poor harvests, price rises had affected the London markets. The price of bread, which in 1767 had reached $8\frac{1}{4}d$ a quartern loaf at Bear Key market, remained high until the late summer of 1768, when improved harvests brought down the price of flour. At the same time that the prices of provisions were high, serious recession was hitting several of the important trades of London, as well as many of the lesser ones. Coalheavers, seamen, silk-weavers, shoemakers, tailors and many other groups were finding employment increasingly difficult to obtain, and their income was too low to meet the rising cost of essentials.

While the problems of most of London's industrious poor had only become acute early in 1768, high prices and unemployment, which had stimulated the provincial hunger riots of 1766, continued to disturb the poor across the country in the following two years. Yet despite these conditions, most of the provincial populace remained passive, while the focus of unrest moved to the metropolitan area. There were some attempts to force down food prices by crowd action, but disorders in the countryside were relatively inconsequential after November 1766. This passivity which succeeded the widespread, violent protests of the summer and autumn of 1766 requires some explanation, for it contrasted with the growing militancy of London industrial mobs.

A variety of influences affected the attitude of the poor outside the London area to their deteriorating conditions. Although to some extent the riots had played themselves out with the arrival of the winter season, vigorous suppression by the army was the immediate cause of the collapse of violent efforts by the country population to reduce food prices and end grain movements to the ports. The early trials of rioters before the special assize courts, and severe sentences of imprisonment, service in the armed forces, transportation or death had their intended effect of cowing the disaffected. Heavily reinforced garrisons in the troubled counties ensured the continued deference to authority. The Ministry's blackmail tactic of holding convicted prisoners under suspended sentences to discourage further protests from their friends and relatives assisted in maintaining the uneasy calm. At the same time, the embargo on all grain exports, and on the use of corn for luxury industries such as starch-making and distillery, together with the duty-free importation of provisions from Ireland, Europe and America, removed the provocation of continued luxury consumption in times of dearth, and complemented the more negative measures of repression. Many of these measures reassured the poor who had initially panicked at the prospect of outright famine.

Although the government's intervention in trade and the provision of subsidised supplies of grain by private associations helped to prevent the prices of food rising to even higher levels, the cost of essentials continued to cause the country populace grave distress. More influential in mollifying the provincial poor's resentment was the reduction of large-scale grain movements, especially to the ports, and the token limitation of the middle-men's activities, following the proclamation of the old statutes against engrossing, regrating and forestalling.

Even more important in the pacification of the countryside was the dawning realisation, confirmed by the second consecutive poor harvest of 1767, that the shortage of food was a natural rather than an artificial one, for which the middlemen could hardly be blamed. The exactions of a 'taxation populaire' no longer seemed appropriate, and indeed had already proved both ineffective and self-defeating. The deprived, therefore, did not resume attacks on middlemen and markets despite continued high prices.

The response of the provincial poor to their difficult circumstances now took several forms. Those with initiative, who had not been incarcerated or impressed, migrated to the home counties, particularly Middlesex and Surrey, or travelled to America, taking their resentment of the social system with them. Others accepted, with varying degrees of gratitude, the subsidised food of private associations of gentry and merchants or of the Ministry. In East Anglia, the violent protests at the building of workhouses for combined parishes died down, as the poor were forced to accept the loss of their ancient freehold, or starve. In general the provincial poor, after their first bitter reaction to the violently fluctuating prices of provisions in the summer and autumn of 1766, sank first into sullen resentment at the military repression, and then into apathetic resignation, as conditions worsened and seemed beyond their power to influence. In a predominantly rustic society, natural scarcity seemed to be divine retribution for past sins, which had to be endured. As the habits of deprivation became familiar once more, the expectations of the poor outside London adjusted more closely to reality.

Far less passive and unsophisticated was the metropolitan poor's response to the coincidence of high prices, and reduced employment in several of the capital's major industries in 1768–9. The economic recession which had earlier caused unrest in the old clothing centres of southern England, the hardware districts of the Midlands, and the mining villages of the north-east now manifested itself in competition for scarce jobs in shipping, on the docks, in silk-weaving and a variety of lesser occupations of the capital. While the provincial poor had become resigned to the decline of their living standards, the metropolitan poor violently reacted to what was a more recent, and therefore novel, downturn. Not yet weakened by deprivation, or dulled by hopelessness, the industrious lower orders of London turned to vigorous industrial protests in 1768.

Before examining in detail the most important industrial riots of 1768–9, some review of their background and consideration of their connection with other serious disorders, which occurred at about the same time, is relevant, for the industrial disputes took place against a backcloth of provincial hunger riots and metropolitan political disorders.

Until now historians have given most attention to the relationship between the Wilkite political riots and the industrial disturbances. The reasons for this are fairly evident. The Wilkite movement's fascination for historians, apparent from the numerous biographies and monographs available, has reflected not only the intrinsic appeal of Wilkes himself, but also, in some instances, Whiggish preoccupation with the evolution of Parliamentary government. The coincidence of industrial disputes and the activities of Wilkite mobs in 1768–9, which seemed related to events in America and Ireland, suggested a conspiracy to contemporaries, who commented upon the causes of English discontents in private correspondence, pamphlets, memoirs and newspapers. Later students of the period accepted the validity of their interpretation of the events.

To one of these scholars, Raymond Postgate, the industrial riots appeared to be the work of Wilkite agitators. Accepting the cries of seamen and others of 'Wilkes and Liberty' and Horace Walpole's dark broodings on conspiracy at their face value, Postgate saw the industrial disturbances of 1768 as political strikes.[1] More recently George Rudé has rejected this interpretation. He sees the two sets of riots following parallel but separate courses.[2]

There is little evidence to support a conspiracy thesis about Wilkite agitators deliberately fomenting industrial unrest. Plainly there was some overlapping of interests. Thus seamen or coalheavers, demanding improved conditions, cheered for Wilkes; or

[1] Walpole was contradictory on the question of political agitators inciting people with industrial grievances. On one occasion he wrote of non-political mobs acting independently. 'We have independent mobs that have nothing to do with Wilkes, and who only take advantage of so favourable a season. The dearness of provisions incites, the hope of increased wages allures, and drink puts them in motion' (Horace Walpole to Sir Horace Mann, 12 May 1768, Walpole's *Letters*, ed. Toynbee, vIII 186–7). Raymond W. Postgate, *That Devil Wilkes* (New York: Vanguard Press, 1929) p. 181.

[2] Rudé, *Wilkes and Liberty*, pp. 94–104. Maccoby denies political exploitation of economic distress but emphasises the connection of political and economic unrest: '. . . the crowds that gathered outside Wilkes's prison became almost as much a school of plebeian economic agitation as of plebeian politics' (Simon Maccoby, *English Radicalism, 1762–1785, The Origins*, vol. I of *The English Radical Tradition, 1763–1914*, ed. Alan Bullock and F. W. Deakin [London: Nicholas Kaye, 1955] pp. 458, 8–9).

political mobs, shouting 'better hanged than starved', displayed a loaf draped in black crape at the Royal Exchange to demonstrate general discontent with the high prices of food. Yet other incidents illustrate the superficial, even contradictory character of the relationship of political and social disturbances. Seamen, having gained some of their economic demands, turned on a Wilkite mob and drove it off. While coalheavers continued to support Wilkes from time to time, their leader Ralph Hodgson voted against Wilkes in the Middlesex elections when many of their chief enemies supported him at the polls.[3] Undoubtedly, disgruntled tradesmen did appear in Wilkite mobs, but they were there as individuals and not as members of an industrial interest which saw itself committed to supporting Wilkite causes.[4] One might suspect that industrial interests were more inclined to exploit the economic grievances of political mobs than political radicals were inclined to stir up the industrious poor for political purposes. The seamen who signed their demands with 'Wilkes and Liberty' were merely using another lever to force the government and their employers to concede their wishes.[5]

In another important sense, the juxtaposition of the industrial and political riots was significant. The fact that there was no evidence of a conspiracy did not prevent the authorities, from the King down to the magistrates, from linking together not only the industrial and political riots in London, but also the English disturbances with events in Boston and Dublin. The nature of the authorities' responses to unrest in the metropolis in 1768–9 can only be appreciated with this consideration in mind. The authorities' tardiness in dealing effectively with the industrial riots was due to their preoccupation with the political riots, which many of them saw as the driving force behind all the disturbances. The stretching of the limited forces of order by the Wilkite demonstrations and riots enabled the industrial riots to gain strength and develop into a greater threat to the social structure than would otherwise have been the case.

Thus one may conclude with George Rudé that neither the

[3] Middlesex Record Office, Westminster, *Middlesex Poll Book* (1768–9) pp. 148, 179.
[4] Rudé, pp. 8–9.
[5] 'Memorials of Dialogues betwixt several seamen, a certain victualler, and a s—l master in the late riot', *Shelburne Papers*, vol. 133, fol. 374.

political nor the industrial riots were subordinate to the other. There was an interaction which was not wholly accidental. Political riots usually occurred in periods of economic unrest.[6] The authorities were inhibited by their distorted perception of events, and thus they permitted disorder to grow before belatedly suppressing it. But in 1768–9 there was certainly no co-odinated plan of political strikes.

The connection between the provincial hunger riots and industrial protests has attracted less comment. Few contemporaries saw the hunger riots as evidence of a conspiracy to overthrow the social order. Until recently, historians found little to interest them in the frequent 'protests of the belly'. Yet there was an interaction between these two forms of social protest which merits closer analysis.

London's remarkable freedom from serious protests over high prices and low wages in 1766–7 was merely a temporary respite from social disorder. The very measures which the ruling interests used to avert a general insurrection in the countryside greatly contributed to rising tensions in the capital. The diversion of the rural poor's hostility towards middlemen and farmers, and away from landowners, had focused attacks upon markets and the movement of food towards the urban centres. Although the central authorities soon forced Lords-Lieutenant and magistrates to safeguard supplies of essential food to the metropolis, and brought in provisions from overseas, food prices inevitably increased in city markets (although at a slower rate than in the country towns). At the same time that military suppression snuffed out radical measures of rural 'self-help' in a time of anticipated famine, the Ministry did nothing to discourage the increasing migration of the destitute to Middlesex and Surrey. Secure from the rigours of the Laws of Settlement, the new arrivals disappeared into the densely populated eastern and southern parishes of London, seeking work in the already overcrowded, unskilled and semi-skilled trades of the capital. This influx of labour as a result of provincial distress in 1768–9 aggravated a crisis already developing from a combination of post-war adjustment, trade realignment and increasingly frequent *laissez-faire* policies of the government.[7]

[6] Rudé, *Wilkes and Liberty*, p. 9.
[7] Rudé, *The Crowd in History*, p. 34.

While at the same time unemployment increased in London, seasonal opportunities for supplementing income declined. Because of the poor harvests of 1766–7, farmers required fewer harvest workers in the home counties.[8] Those Londoners lucky enough to find work on the land were under-employed. Along with their reduced wages, they carried back to the city resentment at the military suppression of 1766. Few were able to reduce the effects of increased unemployment in their London trades by supplementing their income on the land. The cumulative effect of this deprivation of additional income explains why the industrious poor of the metropolis felt acutely the economic recession of 1768–9.

Both those who emigrated to London in search of employment in the 1760s and the Londoners who worked in the harvest fields drew the lesson from the 1766 hunger riots that the traditional response to high prices of trying to force them down was ineffectual. Not only was the enforcing of a 'taxation populaire' unsuccessful in the relatively sophisticated economy of the 1760s, it was self-defeating, for it aggravated the very conditions it sought to alleviate. The authorities might indicate the middlemen as the creators of artificial shortage, but neither they, nor the mobs they encouraged, could end forestalling, engrossing and regrating without starving the population centres.

The tactics of the urban mob now focused on the raising of wages and controlling competition for jobs. Petitions for Parliamentary action to correct wrongs, strikes against employers, internecine struggles for dwindling employment and riots succeeded the earlier provincial attacks upon markets and food in transit. While this urban form of social protest was by no means novel – cloth workers in the West Country, miners in the north-east and Manchester operatives had all struck for higher wages and better working conditions earlier in the century – its extent was new, and contrasted sharply with the form of the provincial riots which closely preceded it.

Clearly separate from the political riots, these industrial disorders of 1768–9 were essentially an extension of the provincial hunger riots in a more sophisticated urban environment. Superficially these industrial disturbances differed from each other, and from the earlier hunger riots. Yet although each outbreak had its

[8] Ashton, *An Economic History of England*, p. 32.

own unique circumstances which modified its shape and direction, these industrial riots and strikes of 1768–9 were reflective of the same type of economic and social ferment that underlay the extensive hunger riots two years earlier. The high price of 'necessaries', too, was the common denominator of all the industrial and hunger protests. In the metropolitan setting the tensions were heightened and the disturbances were more violent.

2 Metropolitan Industrial Disorders

Although disturbances were widespread among the industrious poor of the metropolis in 1768–9, the most serious disorders involved primarily three groups, the coalheavers, seamen and silk-weavers.[1] Because they were in many ways typical of the struggles of the industrious poor of London in the period which preceded the rapidly increasing industrialisation of English society, this chapter will focus upon these three types of disturbances. By describing and analysing the strikes and riots of the coalheavers, seamen and silk-weavers, it will try to show that underlying the precipitating cause of disorder, that of dwindling income in the face of rising prices of 'necessaries', were the tensions of a society in the process of rapid change, which affected the concerns not only of the poor, but also of their employers and the ruling orders. The pre-industrial riots of London, no less than the earlier provincial hunger riots were the product of the interaction of underlying attitudes and ideas of various interests.

I

Because the goals and actions of the London coalheavers, who were the first to engage in extensive disorders in 1768, revealed a mixture of traditional and radical attitudes to industrial relations, they suggested the uncertainties of the poor in a period of rapid change. Broadly the first phase of their responses to worsening conditions of life took the form of customary demonstrations in Stepney Fields, followed by processions to the Palace

---

[1] Other tradesmen, journeymen tailors, coopers and shoemakers, for example, had formidable organisations and rioted in 1768. See Add. MSS., 32990, fol. 77; *Public Advertiser*, 28 May 1768; and *Westminster Journal and Political Miscellany*, 25 June 1768.

of Westminster to present petitions asking for Parliamentary relief, in the fashion characteristic of eighteenth-century protesters.[2] With the failure of these methods, they turned to the more modern technique of creating a tighter organisation along trade union lines. By stopping the flow of coal into the capital, they now sought to pressure their employers into raising their wages to meet the rapidly rising cost of food. When the authorities tried to break their strike by reopening an official registry office and advertising for general labourers from outside the trade, the coalheavers attacked the taverns operated by agents registering the newcomers. Violence reached a peak with the use of seamen to unload colliers in the Thames. The coalheavers made murderous attacks, not only on the seamen themselves, but also upon any who co-operated in the unloading operations. Lightermen who transported the coal from ship to shore, coal-meters who measured the coals landed and many of the general population of Wapping, Stepney and Shadwell parishes lived in terror of armed gangs of coalheavers who roamed the streets in search of strikebreakers. The magistrates only restored calm to the Tower Hamlets after several months of disorders by temporarily garrisoning detachments of Guards in the disaffected parishes, and executing several of the ringleaders for murderous assaults.

For their part, the authorities showed an ignorance of the operation of the economy by vacillating between policies of paternalistic intervention and the abandonment of the coalheavers to the exploitation of a class of middlemen growing affluent upon the developing London market. By indicating clearly to the poor that conditions in the coal trade were wrong and deleterious to their interests, and then failing to correct them, the government stimulated the coalheavers to take matters into their own hands. Far from calming these labourers, the government agitated them further, in much the same way as they had earlier encouraged the provincial poor to regard the middlemen of the provisions trade as responsible for the food scarcity and thereby gave direction to the food riots.

Precipitated by what the coalheavers saw as attempts to reduce their wages and put them back in the power of the undertakers, the coalheavers' disturbances of 1768 took place against a back-

[2] *Annual Register*, xi (1768) 108–9.

ground of rising food prices, a serious disruption of overseas trade and general industrial unrest. But they had a long and complicated history of misery and exploitation, which stretched back into the sixteenth century. In order to understand how the position of these labourers deteriorated seriously in the 1760s, and the nature of their response to their aggravated problems, it will be necessary to go back a few years, at least as far as 1758, when an Act of Parliament, 31 George II, cap. 76, was passed for their relief.

The preamble of this act blamed the problem of the coal trade in London on the absence of an 'established method of hiring, employing and paying' the coalheavers, which enabled the undertakers to oppress them. The trade operated as follows.[3] Having agreed with the master of a collier to unload his ship, these middlemen hired gangs consisting of about fourteen labourers. The undertakers, or their agents, were invariably tavern-keepers, who showed preference to those men who frequented their premises and ran up heavy drinking debts.[4] Although they were paid a rate of pay considerably in excess of what other London labourers received, owing to the heavy manual work they had to perform, the coalheavers never took home more than a fraction of their earnings, because of the many deductions for items such as liquor consumption in the tavern and on board ship, commissions for ships' masters and rent for tools. One newspaper claimed that coalheavers earned 8/- or 9/- per day at a medium of 20d per score of London chauldrons unloaded, but they received only 9d per score after various deductions.[5] It was the arrogant assertion of their monopoly over the marketing of coal shovels, however, and the extortionate charges they made for renting them, which brought the undertakers into conflict with the Commons in

---

[3] 'The Present State of the Coalheavers, Explained and Considered' (30 June 1768), *Shelburne Papers*, vol. 130, fol. 104.

[4] Jonas Hanway, *Letters*, xxxi, reported men who earned 20/- on a long summer's day drank 15/- or 16/- in beer. See also 'Minutes of Evidence Taken before a Select Committee on Drunkenness' (printed 5 August 1834), Add. MSS., 27830, fol. 72; and Patrick Colquhoun, *Treatise on the Commerce and the Police of the River Thames* (London: Joseph Mawman, 1800).

[5] *Lloyd's Evening Post*, 13–15 June 1768. See also George, *London Life in the Eighteenth Century*, pp. 166, 397.

1758 and resulted in the passage of a regulatory act to protect the exploited labourers.[6]

The main intent of this act was to place the coalheavers under the direction of the Alderman of Billingsgate Ward, with whom they were to register their names. For his part, the Alderman had to organise gangs of coalheavers to unload the ships of those masters who applied to him. The Alderman, or his deputy, was to collect the wages of the labourers from the collier captains and, after deducting 2/- in the pound to defray the expenses of operating the registry and to provide a fund for the payment of sickness, burial and dependents' benefits, he was to divide up the remainder among the men. There was also a clause in the act which prohibited anyone from retaining any further part of the coalheavers' wages under a penalty of £50. In practice the authorities did not enforce this provision; nor did Parliament enact another clause which provided for wage-fixing machinery.[7]

The provisions of this act reflect some of the traditional goals of the coalheavers.[8] Ideally they wished to be a fellowship or corporation comparable to the Fellowship Porters, which had originally controlled the unloading of coal, before it restricted itself to cleaner cargoes.[9] Frustrated in this ambition by successive governments, the coalheavers pressed for a regulation of their wage rates, the creation of a benefit fund to insulate them against the disasters of sickness and injury to which their heavy work rendered them prone, and Parliamentary protection from the exploitation of coal-undertakers. In these demands they were consistent with other London tradesmen who did not enjoy the protection of incorporation, as they were in the use of the common eighteenth-century technique of petitioning for relief through private acts of Parliament. By these methods they sought to prevent the erosion of their living standards, rather than to improve them radically.[10] In this they displayed the essentially conservative goals of the industrial poor who rioted over wages and con-

---

[6] M. Dorothy George, 'The London Coalheavers: Attempts to Regulate Waterside Labour in the Eighteenth and Nineteenth Centuries', *Economic History, Supplement of Economic Journal*, I, no. 4 (1926–9) 229–48.

[7] 'The Present State of the Coalheavers', *Shelburne Papers*, vol. 130, fol. 111.

[8] George, 'The London Coalheavers', p. 234.

[9] British Museum, 'The Coalheavers' Case' (1764).

[10] *Westminster Journal and London Political Miscellany*, 21 May 1768.

ditions in the 1760s. It was rather in their tactics, than in their goals, that they became radical in 1768.

The evident deficiencies of the Act of 1758, on the other hand, suggest a qualified attitude towards industrial regulation on the part of the governing orders. These deficiencies were threefold. First, and most important, the legislation was permissive. The coalheavers were free to register or not with the official registry as they wished. Similarly, ships' masters were not obliged to hire through that office. This left the undertakers, who were influential amongst shipowners, free to destroy the coalheavers' confidence in the official scheme, and eventually to trick the illiterate labourers into accepting once again their control of the trade. Second, a wage-fixing clause was omitted. As one critic noted:

The want of this power defeats some other provisions in the act, for though the alderman or his deputy may appoint a gang of labourers to go to work on board any ship, and the labourers are subjected to a penalty if they refuse or neglect to obey such an order, yet no terms being settled if the wages offered should not be agreeable to them and on that account only they refuse to work no penalty could be enforced nor the law have any affect on them.[11]

But the omission threatened the interests of the men rather than those of their employers. Without wage-fixing machinery the coalheavers were left to the mercies of scheming undertakers and the vagaries of the North Sea weather, which created fluctuations in wages by interrupting the regular arrival of coal vessels in the port of London. Third, there was no provision made for the proper enforcement of clauses which forbade any deductions (other than the 2/- in the pound by the official registry) through the establishment of an effective inspection scheme.

Despite these shortcomings, the Act of 1758 relieved the distress of the coalheavers for a while, because it warned the undertakers of the danger of antagonising Parliament further. The men registered their names with the Alderman's deputy, Francis Reynolds, who had been the prime mover for government intervention, and they built up a considerable sum in the benefit fund from the 2/- in the pound deductions. But in time matters played into the hands of the undertakers. The scheme fell into disrepute

[11] 'The Present State of the Coalheavers', *Shelburne Papers*, vol. 130, fol. 111.

when it was discovered that Reynolds had either embezzled the money in the fund, or at least was unable to account satisfactorily for its disposal. Whereupon the undertakers, who had been biding their time, were able to denounce the system of deductions, and, because of the permissive character of the act, to bring the labourers back under their control.[12]

In the 1760s, economic developments operated against the interests of these impoverished coalheavers, and aided the undertakers in their campaign to tame them. The continued growth of London made the metropolis more than ever dependent upon the smooth flow of fuel to the consumers. Thus the government was unwilling to endanger this trade by intervening too vigorously on behalf of a minority group, consisting in the main of dangerously turbulent, Roman Catholic Irishmen, unless serious disorders occurred.[13] Technological change, too, after the mid-century rendered the coalheavers more vulnerable to exploitation. Hitherto the unloading of colliers had been performed by gangs of labourers, who shovelled the coal on to platforms rigged progressively up the sides of the holds. This rather primitive method of operation placed a premium, not only on great physical strength, for which the Irish were famed, but also on experience and skill. The introduction in 1758 of a method of 'whipping' baskets of coal from the holds by means of pulleys made the coalheavers vulnerable to the influx of unskilled labour. By 1768, one-third of all coal was unloaded in London by 'whipping'.[14] Exploiting this development, the undertakers brought on to the London docks labourers from Ireland, Scotland and other parts of England, with the promise of high pay. The labour surplus they created in this way aggravated an existing problem which had been produced by the demobilisation of large numbers of seamen and a post-war depression. The crux of the problem for

[12] Ibid., fol. 110–11. See also *Treasury Solicitor's Papers*, T.S. 11/443/1408; and George, 'The London Coalheavers', p. 236.

[13] A Crown brief estimated 670 men in the coalheaving trade in 1768, of whom two-thirds were Irish Roman Catholics, 'with 70–100 the very dregs of mankind, capable of all kinds of mischief and barbarities living in Wapping, Shadwell and the neighbourhood, where they are encouraged in their violence, and of late, since they refused to work, supported by a set of publicans mostly Irish Roman Catholics' (*Treasury Solicitor's Papers*), T.S. 11/818/2696).

[14] George, 'The London Coalheavers', p. 229.

the coalheavers in their struggle against the undertakers was the ready availability of alternative sources of labour. Not surprisingly the most violent disorders of 1768 were those caused by seamen who broke the coalheavers' strike by unloading their own ships.

The decline of wages of coalheavers in the 1760s, quite apart from the increasing number of deductions, was a reflection of the growing desire of undertakers to exploit this labour surplus and reduce their operating costs. In 1756 the highest rate had been 3/- and the lowest 1/- per score of London chauldrons unloaded; during the following year the rate varied from 1/- to 2/9 per score; but by 1768 the employers regarded anything over 1/6 per score excessive, and anything over 2/- was quite exceptional.[15] The conditions of coalheavers, therefore, had greatly deteriorated owing not only to the rising cost of food, but also to the reduction of their wages. In contrast to their agricultural counterparts, the coal-labourers were more conscious of the decline in wages than the rise in the cost of bread. The deep-seated grievances they harboured against the undertakers diverted their animosity away from middlemen of the provisions trade towards their employers, who were able to redirect it first towards the agents of the official agency and then towards rival labour interests.

The Alderman of Billingsgate Ward, William Beckford, now lost interest in operating an official registry. The coalheavers were more than ever in the power of their old enemies, the undertakers, who found new excuses for deductions from their wages despite the legal penalties. A petition of the coalheavers to amend the 1758 Act to make it more effective failed in 1764.[16] Thereafter conditions in the coal trade deteriorated until the labourers found them intolerable in 1768.[17]

In 1765 the violent strike of pitmen on the rivers Tyne and Wear, which seriously interrupted the flow of coal to the capital, raised the price of coal, and reduced employment on the London docks, provided a prelude to the metropolitan coalheavers' riots

---

[15] Ibid., p. 231.

[16] 'The Coalheavers' Case' (1764).

[17] 'Representation of the Justices for Middlesex in General Session Assembled at Hicks Hall' (September 1768), Middlesex Record Office, Westminster, *Session Papers*, refers to coalheavers' strike about the beginning of February over wages.

of 1768. The northern miners demanded an end to a form of bond servitude and a seventy-five per cent increase in their wages of 12/- to 14/- per week. These demands were the response of some provincial industrial workers to the high prices of food which elsewhere in rural England stimulated the widespread hunger riots in 1765 and 1766. Despite the presence of troops, the rioters destroyed colliery equipment and fired coal above and below the ground. Disorders lasted from 14 August to 2 October when the colliery owners agreed to a settlement, the immediate result of which was that the price of coal in the port of London which had risen steeply during the strike fell 32/- per chauldron and coalheavers returned to work as the colliers began to arrive once more. This strike in the north of England had a twofold effect upon the London coalheavers: it aggravated for a time their problems and its apparent success encouraged them to anticipate similar success from direct industrial action. It is also an example of violence beginning in the provinces and later spreading to the metropolis, which was a significant aspect of many disorders in the 1760s.

The first phase of the coalheavers' campaign to improve their conditions in 1768 was the establishment of a system of licensing workers in their trade. The 'market men', who were the landlords of the taverns where men were hired, now agreed with coalheavers to a rate of 20*d* per score of London chauldrons unloaded, with 6*d* deducted for a benefit fun.[18] Organised by Ralph Hodgson, a justice of the peace in the Tower Hamlets, this scheme involved the establishment of a private agency in opposition to the undertakers.[19] Its method of operation along trade-union lines represented a new phase in the coalheavers' long struggle against their exploiters. As a measure of self-help in the face of the collapse of the government-sponsored scheme, it was in distinct contrast to their former reliance upon government paternalism. Hodgson encouraged their sense of identity by

[18] *Lloyd's Evening Post*, 13–15 June 1768.
[19] The *Session Papers* blamed Hodgson for 'notorious inactivity and supineness, with regard to any measures for quelling the riots . . . many factious and inflammatory advertisements and paragraphs . . . in the public newspapers'. The justices regarded him 'as accessory to outrages and disorders aforesaid to the great disturbance of the public peace and terror of his majesty's subjects' ('Representation of the Justices for Middlesex', *Session Papers*).

publicly patronising the coalheavers' festivals, especially the St Patrick's Day parade. He later denied the 'rabble-rousing' accusation and claimed his patronage was to avert disorders: 'It would have served them to rivet the false and groundless report they have propagated of my being an Irishman; a report, however, which I shall never rank in the list of their scandals. Since if there is a more than ordinary mean, dirty, illiberal injustice, it must be that of national reflections.'[20] Hodgson organised the men into a society, the Bucks, which had forty-five governors, and which met at the Sign of the Horse and Dray in New Gravel Lane, and later at the Swan, King James's Stairs.[21] An Irish terrorist gang, the Whiteboys, which had recently been driven out of Ireland, provided the disciplined core of the organisation in much the same fashion as militiamen and veterans of the Seven Years' War did in the rural hunger riots in 1766.[22] Oaths of secrecy exacted from all members of the Bucks strengthened its unity, if they made the society more suspect in the eyes of the authorities. Justice Hodgson and his clerk, Dunster, charged a fee of 6d for registry in the 'union', and issued all members with a ticket certifying their qualifications to follow the coalheaving trade. The Bucks enforced the closed-shop principle by ducking reluctant labourers in the Thames on the end of a rope until they agreed to join. During the coalheavers' strikes, which began in early 1768, pickets roved the waterfront persuading seamen, coalmeters, carters and others to refuse to handle the coal aboard the anchored colliers.[23]

Now the activities of Hodgson and the coalheavers attracted the attention of William Beckford, who required his agent, Russell, to reopen the official registry. Coalheavers were persuaded to register by the prospect of a rate of 2/- per score and they agreed to 2/- in the pound deductions. Disorders later arose out of the collier masters' refusal of these rates and their use of seamen to unload the ships.[24] The rekindled interest of Beckford

[20] Lloyd's Evening Post, 26–9 August 1768.
[21] Treasury Solicitor's Papers, T.S. 11/818/2696.
[22] Walpole to Sir Horace Mann, 22 June 1765, Walpole's Letters, ed. Toynbee,, VII 203. See also Correspondence of King George the Third, ed. Fortescue, pp. 310–28; Treasury Solicitor's Papers, T.S. 11/818/2696; and Walpole, Memoirs of the Reign of King George the Third, p. 148.
[23] George, 'The London Coalheavers', p. 237.
[24] Lloyd's Evening Post, 26–29 August 1768.

is remarkable. In this he may have acted of his own volition to break the strike, he may have responded to pressure from the Ministry concerned with the disorders and the paralysis of trade, or he may have acceded to the demands of merchants, factors or shipowners whose interests were threatened. Whatever his motivation, in reopening the registry Beckford was doing something that the undertakers themselves welcomed, for they were happy to see the coalheavers' confusion over rival registries. If sufficient violence could be provoked, the undertakers could expect military suppression of the coalheavers and future government indifference to the plight of the 'wilful' coalheavers. The rumour that the coalheavers were receiving encouragement and help from some quarter may well have had some basis in fact, for the lengthy disruption of work meant destitution to the families of coalheavers without financial support from somewhere. This support could have come from undertakers anxious to foster violence in order to restore their control over the trade.

Even more extraordinary than Beckford's rekindled interest was the action of his agent, Russell. He appointed as his clerks two former coal undertakers and victuallers, John Green and Thomas Metcalfe. Whether this was accidental or by design is hard to tell. Both men had experience in hiring practices in the trade but their former associations were fatal to the success of the operation. The coalheavers interpreted the moves of Beckford and Russell as attempts to put them back in the power of the undertakers and to break their strike with the use of cheap 'blackleg' labour.[25]

The subsequent disorders, which were the first of many serious industrial disturbances of 1768, were directed against the taverns of Russell's clerks, where advertisements for general labourers from outside the coalheaving trade were posted. On two occasions in late February 1768 rioters attacked Metcalfe's Salutation Inn, Wapping Wall. They 'pulled down the chimney piece, broke the windows, china bowls, decanters, and almost everything else in the bar' and threatened to murder Metcalfe, who had wisely if ungallantly fled leaving his wife to face the infuriated coalheavers.[26] Thereafter in March and April riotous mobs attacked

[25] 'The Present State of the Coalheavers', *Shelburne Papers*, vol. 130, fol. 106.
[26] 'Petition for Redress by the Inhabitants of St. Pauls, Shadwell and Adjacent Places in the County of Middlesex' (19 May 1768), Middlesex Record Office, Westminster, *Session Papers*.

the home of Burford Camphire, a leading undertaker, and destroyed property.[27] A more bloody and sustained riot occurred late in April when the coalheavers besieged John Green's Roundabout Tavern for over thirteen hours. During this affair, considerable property damage was done. One account described the walls and ceilings being 'riddled with bullets, and three barrowloads of brickbats and stones were taken out'. Seven heavers were later executed and three transported for their share in this affray.[28] Green himself narrowly escaped with his life.[29] Two deaths occurred during these disorders, and later as an act of revenge the rioters tore to pieces Green's sister as she was about to celebrate her brother's deliverance.[30]

Hodgson certainly encouraged the coalheavers in their opposition to Beckford's registry. He persuaded them that a deduction of 2/- in the pound was 'insupportable' at the current rates of wages. As a magistrate, he made little effort to suppress their disorders. His clerk, Dunster, was wounded in the hand during the siege of the Roundabout Tavern. When Hodgson himself belatedly arrived on the scene, he arrested Green and his confederate, a seaman who happened to be in the house when the attack began and who elected to help Green defend himself, on a charge of murder, while he permitted their assailants to disperse unharmed.[31]

A few days after the attack on Green's premises, the Wapping coalheavers turned their attention to putting pressure on both Parliament and their employers for relief. Reportedly some of them petitioned Parliament, while others, 'complaining of low wages, liquor payments, and bad quality goods', stopped colliers working in the Thames. Large groups of coalheavers followed up these activities by visits to employers, whom they frightened into promising higher wages. Then dissatisfied with oral promises given under duress, they tried unsuccessfully to persuade Thomas

---

[27] Camphire was plainly a man of substance, for the mob broke into his coach-house and cut the harness of his chariot.

[28] St. James's Chronicle, 7–9 July 1768.

[29] George, 'The London Coalheavers', p. 238.

[30] Walpole's Letters, ii 208.

[31] 'The Present State of the Coalheavers', Shelburne Papers, vol. 130, fol. 106.

Harley, as Lord Mayor, to intervene to ensure that the employers lived up to their word.[32]

A more promising turn of events came with a meeting in Whitechapel of the foremen of the labouring gangs, a few owners of coal vessels, and some justices of the peace, at which complaints of low wages and the 2/- deductions required under the 1758 Act were discussed. The outcome of this gathering was an agreement that wages would be raised to 2/- per score of London chauldrons of coal unloaded. This figure was almost 6d more than the current medium rate for the year round. (Wages varied with the season and the number of colliers to be unloaded on any given day.) At this rate of pay, the coalheavers had no objection to paying the 2/- in the pound levy and appeared to accept the reopening of Beckford's agency. Some owners paid the 2/- per score required, owing to the arrival of a large fleet of colliers about 20 May.[33] But unfortunately the majority of collier owners, who did not attend the meeting, did not consider the agreement binding on them. They claimed they were unable to pay the new rate, and some of them agreed with their crews to unload the coals without the aid of coalheavers.[34]

The infuriated labourers now resumed their strike, stopping work on the wharfs from Limehouse to Westminster. The heavers stopped West Country and other barges from unloading 'and every person they found doing the least business they made ride (as they term it) the Wooden Horse, which is carrying them on a sharp stick, and flogging them with ropes, sticks, and other weapons'.[35] They tried to block all coal movements in the metropolis, even to the extent of unharnessing horses from coal carts in the Strand. Particular objects of their fury were the seamen, together with the lightermen and metermen who co-operated in the work of unloading.[36] Although the acquittal of Green on a charge of murder exasperated them, the use of seamen as strike-breakers diverted their attention from Beckford's agent and the undertakers. Reportedly seamen now earned 6/- a day at the rate

[32] *Annual Register*, XI (1768) 101–2; and *Gentleman's Magazine*, XXXVIII (1768) 243, cited by Rudé, *Wilkes and Liberty*, p. 98.

[33] *Treasury Solicitor's Papers*, T.S. 11/818/2696.

[34] 'The Present State of the Coalheavers', *Shelburne Papers*, vol. 130, fol. 107–8.

[35] *St. James's Chronicle*, 14–17 May 1768.

[36] Ibid., 28–31 May 1768. *Lloyd's Evening Post*, 13–15 June 1768.

coalheavers refused to work.[37] The authorities responded by advertising for 'three hundred men, or more, as coal-fillers and carmen, whose wages will be a guinea a week and upwards' and promised them protection.[38] There were now bloody riots between collier seamen and the coaheavers.[39] The latter kept a twenty-four-hour watch on the landing steps of the Thames, in an effort to starve out the seamen.[40] They attacked colliermen who ventured ashore, and in one bloody affray they brutally murdered a young seaman, John Beatty.[41] Seamen were obliged to keep close watch against night attacks, and for several nights there was continual firing between ship and shore.[42] Matters reached a crisis at the end of May with a 'numerous meeting of seamen on Newcastle and Sunderland colliers' which proposed to offer to unload all colliers.[43] Bloody riots continued, and a state of near anarchy existed in Wapping and neighbouring parishes during May and early June. Despite their relatively few numbers of about seven hundred, because of their unity and violence the coalheavers had a serious effect on the operation of the docks. Their actions in the early summer of 1768 were more formidable because they coincided with other industrial disputes and extended further the forces of order. Only the intervention of the Guards and their temporary stationing in Wapping and Shadwell pacified the coalheavers, who now returned to work at the old figure of 1/6 per score.[44]

Despite some short-term gains, the coalheavers were not successful in permanently relieving themselves of oppression by their blend of old and new tactics. Although the official registry opened in July 1768 after the execution of nine of their ringleaders, men who registered there were not employed by the undertakers if they could find others.[45] In 1770, a more effective act replaced the 1758 Act. Under it, the undertakers were prohibited from

[37] Lloyd's Evening Post, 13–15 June 1768.
[38] St. James's Chronicle, 17–19 May 1768.
[39] Ibid., 24–26 May 1768.
[40] Universal Magazine, XLIII (May 1768) 330.
[41] Treasury Solicitor's Papers, T.S. 11/818/2696; Public Advertiser, 21 July and 8 August 1768, cited by George, 'The London Coalheavers', p. 239.
[42] St. James's Chronicle, 2–4 June 1768.
[43] Public Advertiser, 30 May 1768.
[44] St. James's Chronicle, 17–19 June 1768; and Public Advertiser, 17 June 1768.
[45] Universal Magazine, XLIII (27 July 1768) 53.

being victuallers, deducting amounts from the coalheavers' wages or paying in other than current coin. A mechanism for settling disputes in the trade which involved the Mayor and Aldermen was established. This group meeting in January of each year could adjust the fixed rate of 1/6 per score should circumstances warrant it. Captains could however employ their own men at rates mutually agreed upon. While there were obvious weaknesses in this legislation which tended to erode the coalheavers' position, it worked well for three years, after which the authorities permitted it to lapse.[46]

A consideration of how this group of middlemen, i.e. the undertakers, managed to survive despite their dubious economic value, frequent disruption in the trade and the evident hostility of the authorities, reveals a confusion of attitudes and goals of the ruling orders in a period of rapid economic and social change which paralleled that of the industrious poor themselves, and contributed to the social tensions.

As noted earlier, the activities of these undertakers had first attracted the serious concern of the government in 1758, at a time when middlemen were coming under heavy attack in the press and in pamphlets for their monopolistic activities, which their critics believed raised the price of provisions and caused dangerous unrest among the poor. The revelation of many of the undertakers' extortions generated some sympathy for the hapless coalheavers among the ruling orders who resented the conspicuous consumption of yet another group of rising middlemen. One newspaper wrote of the undertakers' aping their betters.[47] But it was not so much the periodical petitions of the coalheavers as the discovery that undertakers were acting in restraint of trade by engrossing the coal-shovel supply that galvanised the government into attempts to regulate the coalheaving trade. An insolent challenge to Parliamentary privilege from one of the leading undertakers provided a final stimulus for official action to relieve the coalheavers. Yet the subsequent legislation was ineffective in restoring some balance to the coalheaving trade.

Superficially this failure of Parliament was surprising because the role of the undertakers was not only unproductive but also

[46] Colquhoun, *Treatise on the Commerce and the Police of the River Thames*, pp. 144 *et seq.*
[47] *Westminster Journal and London Political Miscellany*, 21 May 1768.

unique in the hiring practices of the 1760s, according to one adviser of the Ministry.[48] In no other 'laborious handicraft' were intermediaries involved directly in the hiring of labour. When a master tailor, shoemaker or weaver required hands, he merely went to an inn known to be frequented by journeymen of his trade. Even waterfront workers most comparable to coalheavers, such as lumpers who unloaded East Indian and American vessels, and ballast heavers, gathered at 'houses of call' where mates went to hire them. There was nothing in the coalheaving trade that made middlemen essential in the hiring process. Why then did the Ministry not sponsor a more drastic measure which would have permanently destroyed the influence of an interest which many believed was responsible for increased prices of one of the 'necessaries' of life? One important reason for the moderation of the government was the great influence of the undertakers.

This influence had developed gradually over a lengthy period for various reasons. Because the unloading of coal ships was one of the least attractive of London's occupations, it was left to recent immigrants. The very heavy nature of the work gave Irishmen a natural advantage over the less muscular, native-born Londoners. Because of their illiteracy and ignorance of urban life, Irish immigrants quickly fell victim to the scheming undertakers. Growing wealthy upon their extortions, the undertakers by the mid-century had bought into coastwise shipping.[49] In doing this, they further encouraged the interrelatedness of provincial and metropolitan industrial disputes. They were now able to influence the hiring practices of collier captains.[50] Coalheavers who alienated the undertakers soon found it impossible to follow their trade. If they insulted or obstructed these middlemen, the magistrates, amongst whom the undertakers had gained great influence, used the power of the law to restrain them. Many of the headboroughs, who arrested the coalheavers on spurious charges, were often victuallers, who were themselves undertakers or their friends.[51] Several of the magistrates most active in per-

[48] 'The Present State of the Coalheavers', *Shelburne Papers*, vol. 130, fol. 109.

[49] George, 'The London Coalheavers', p. 234.

[50] 'The Present State of the Coalheavers', *Shelburne Papers*, vol. 130, fol. 114.

[51] British Museum, 'The Case of Mr. Francis Reynolds', p. 1; and *Treasury Solicitor's Papers*, T.S. 11/818/2696.

secuting the coalheavers or their leaders were leading undertakers such as Burford Camphire, or the sons of coal-buyers, such as John Shakespeare. These justices used their influence to deny a victualling licence to Francis Reynolds, an Irish innkeeper who led the coalheavers' campaign for government intervention in 1758. As a result of this discrimination, Reynolds was forced into bankruptcy and finally into the parish workhouse in Shadwell.[52] In 1768 the limited support for a petition to remove a magistrate sympathetic towards the heavers suggests the distaste of the more able justices for the undertakers' vested interests on the bench. At the Middlesex General Quarter and General Sessions of the Peace, which discussed Hodgson's role in the coalheavers' riots, only a minority of justices voted in favour of the removal of his name from the Commission of the Peace for the county (six out of fourteen). Of these at least two, Shakespeare and Mainwaring, were notorious for their connection with the undertakers. Even more impressive was the calibre of the eight abstaining in the voting. John Hawkins, Chairman of the Sessions, Robert Pell and Saunders Welch were among these.[53] Nor did the influence of the undertakers end with the local authorities; it extended into Parliament itself. The appointment of Justice Shakespeare to be one of two inspectors of coal-meters, a very sensitive post in the struggle between coalheavers and undertakers during the strikes of 1768, illustrates this influence.[54]

Faced with this entrenchment of the undertakers, there were broadly two approaches to the problems of the coalheaving trade that the Ministry could adopt, once they had decided to intervene. Either they could apply rigorously the same paternalistic principles behind the old anti-middlemen statutes concerned with the provisions trade, the assize of bread and the statute of artificers, or they could implement the newer concepts of freeing trade from 'unfair combinations' which diverted it from its 'natural channels', and leave it alone to prosper with a minimum of interference.

---

[52] 'Case of Mr. Francis Reynolds'.

[53] Middlesex Record Office, Westminster, *Order Book*, Middlesex General Quarter and General Sessions of the Peace, 3–14 George III, no. 8.

[54] One newspaper reported two offices to be established under the inspection of Alderman Beckford and Justice Shakespeare at Billingsgate and Shadwell (*Lloyd's Evening Post*, 1–3 June 1768).

Translated into specific actions, this would have meant, in the first instance, obliging the coalheavers to register at Beckford's official agency, whose monopoly in the provision of labour the collier captains would observe under heavy penalties. Implicit in such a scheme was the provision of some machinery for the setting of wages acceptable to both the labourers and the shipowners. Such action would have destroyed the undertakers as an interest in the coalheaving trade, if it were consistently applied. But it would have been a backward-looking step, in the sense that there was already occurring a progressive abandonment of regulations against middlemen of the provision trade, and the assizes of bread and wages. Certainly it conflicted with not only the ideas of an influential minority of intellectuals who favoured freer trade, but, more important, with the realities of economic growth and bureaucratic inadequacies. In making such a scheme of regulation effective, one critic noted, the government would have had to place dangerous powers in the hands of the official agency. Such powers would have been threefold. First, the necessary discretionary power to register those who applied to the registry and to remove from the register those who misbehaved was very liable to abuse. Second, the power to appoint gangs to ships meant that when there were many gangs and few ships to unload there was an opportunity to favour some men; and when there were many ships to unload and few gangs unemployed there was an opportunity to show favour to some masters. Thus there would be opportunity for bribery. Third, if there was power to alter the rates of wages, there would have been a great field for abuse; while on the other hand, fixed wages could be oppressive to men or employers according to circumstances.[55]

The second, broad approach might have resulted in the following. Parliament would have passed an act requiring masters, or their mates, to hire coalheavers directly, probably from 'houses of call' like those frequented by other London trades, and agree with them over wages, which would have been paid on board ship without deductions. For their part, the coalheavers would have had to complete the unloading of any ship once begun, upon penalty of forfeiture of wages in case of default. These measures

---

[55] 'The Present State of the Coalheavers', *Shelburne Papers*, vol. 130, fol. 111–12.

would have removed the need for private and official middlemen. Although such regulations had the advantage of simplicity, there were evident weaknesses which demand only cursory attention here. The greatest problem would have been policing the trade without an effective inspectorate, in the face of the undertakers' entrenched position.

In the event, the course chosen by Parliament was a compromise between the two extremes. The establishment of an official registry was an understandable, if unwise, return to the older, more familiar principles of paternalism. But the permissive character of the 1758 Act represented a recognition of the newer principles of *laissez-faire*. The subsequent amending act of 10 George III, cap. 53, which forbade victuallers' engagement in undertaking or the payment in other than coins of the realm free from any deductions, and fixed wage rates at 1/6 per score of London chauldrons (with provision for the Mayor and Aldermen to settle disputes and appeals to Quarter Sessions) was a piecemeal treatment of a problem that required more thoroughgoing action to combat the power and influence of the undertakers, and was also permissive.

Successive Ministries were torn between their antipathy towards middlemen of all types, who created dangerous unrest among the lower orders, and the increasingly popular concept that trade flourished best when it was free from outside interference. They therefore imposed ineffectual remedies. Indeed they added to the chaos of the trade. By introducing an official registry alongside the private system of hiring provided by the undertakers they heartened the coalheavers for a while. When the government temporarily withdrew from the field, they encouraged the emergence of a third system of hiring sponsored by Justice Hodgson. With the operation of all three systems at a later date, they confused the coalheavers, who were 'cast from one to another'. Far from correcting a deplorable situation, the government made it worse by stimulating rivalry among the coalheavers who supported different hiring systems. By encouraging the poor to believe that improvement in conditions was possible, and then failing to deliver that improvement – always a dangerous course to follow in the eighteenth century – the government encouraged the coalheavers to take matters into their own hands. Thus the government's intervention in the coalheaving trade, like its attempt to control the middlemen of the provision trade, stimulated rather than discouraged disorders.

Between 1758 and 1773, despite the actions of the government to regulate more closely certain facets of economic life, the non-interventionists' influence grew. This was due not only to the persuasiveness of their arguments, but also to economic realities. The increased sophistication of the economy demanded the multiplication of middlemen. At the same time there was growing disillusionment with government intervention, stemming from the experience of the provincial riots when attempts to regulate provision merchants and others had aggravated rather than solved problems. Perhaps the final straw which led eventually to the abandonment of the coalheavers to their fate was the fear created by the widespread agricultural, pre-industrial and political riots of the 1760s. Frightened by the general disaffection of the lower orders, the government came to see the Irish coalheavers as a dangerously turbulent group whose power to disrupt the waterfront was out of all proportion to their numbers, for they were able to stimulate other more numerous and essential groups, such as seamen, to riot and strike. The undertakers might be anathema to the ruling orders who disliked their brash, conspicuous consumption, but they promised to fulfil a valuable role in controlling one of the most turbulent of London trades. The length and violence of the coalheavers' disorders in 1768 alienated the ruling classes despite their initial sympathy for the labourers' depressed conditions. The sensitivity of the government to the threat to the peace of the Tower Hamlets and other parishes in the east and south of London is apparent in their constant search for suitable magistrates who would live in their jurisdictions and be readily available in times of civil commotion. The pensioning of John Green, who attracted attention for the first time by his resolute defence of the Roundabout Tavern in 1768, and his subsequent appointment to the Middlesex bench despite his inferior social background, illustrate this concern.[56] In contrast, the demise of Ralph Hodgson as magistrate despite his gentry status

---

[56] Undated memo endorsed by Charles Townsend 'to appoint Mr. Green, Justice of the Peace for Westminster—no salary expected as he already enjoys a pension of £280 per annum for former services' (*Sydney Papers*). See also *Shelburne Papers*, vol. 124, fol. 131, which records annual pension to John Green of £200; and Norris, *Shelburne and Reform*, p. 193.

and his membership of Gray's Inn and the Inner Temple, for his share in encouraging the coalheavers to challenge authority, indicates official response to apostates in the ruling orders.[57]

The abandonment of government protection for the coalheavers after 1773 occurred at the same time that the anti-middlemen statutes were repealed, the Pownall Act revised upwards the thresholds at which corn bounties were to be paid, and the bread and wage assizes were in the course of abandonment. This was not coincidental. Yet the decisive reason for the lapse of protection in the coal trade was the fear of the disruptive capacity of the coalheavers among the London poor, and the expectation that if left to their own devices the undertakers would control this dangerous element in lower-class society, even if in the process they grew rich on their extortions.

II

While the high price of food and the deepening depression in overseas trade brought matters to a crisis in the spring of 1768, the underlying causes of unrest among seamen, as among coalheavers, had a lengthy history. Once again it will be necessary to examine several years of background to the strikes and demonstrations, which occurred between early May and late August 1768 in order to understand the motives of the strikers, and the responses of their employers and of the authorities.

After the peace settlement of 1763, seamen faced a severe deterioration in their circumstances.[58] Because this downswing followed several years of prosperity and expanded employment (the first three years of the war had seen a decline in the numbers of merchant seamen, but in the last four years there was a general expansion of labour up to a peak in 1763), the expectations of seamen were at considerable variance with their economic circumstances.[59] Not only were they dissatisfied with their standard of living, mariners also found their peacetime role disappointing. They were no longer the heroes of the nation. With reason, they

---

[57] *Pension Book of Grays Inn, Records of the Honourable Society, 1669–1800*, ed. Reginald J. Fletcher, B.D. (London: Chiswick Press, 1910) p. 267.

[58] Rudé, *Wilkes and Liberty*, p. 91.

[59] Ralph Davis, 'Seamen's Sixpence: An Index of Commercial Activity, 1607–1828', *Economica*, new ser., XXIII (1956) 328 *et seq.*

felt their countrymen had short memories indeed. Many summed up their complaints in the following terms: 'We, seamen have been slighted and our wages reduced so low and provisions so dear that we have been rendered uncapable of procuring the common necessaries of life for ourselves and our families.'[60]

The most immediate economic problems, which derived from the decline in employment opportunities, were related to a number of causes. The first of these was a drastic reduction in the size of the army and navy. In little more than a year, 153,000 men were demobilised.[61] While this demobilisation was most rapid in 1763 and 1764, it continued in subsequent years, although at a reduced pace. Thus between January and September 1767 the number of men borne in the Royal Navy dropped from 17,694 to 15,280.[62] At a time when many of these veterans sought civilian jobs, significant reductions in the merchant service occurred. Not only was there an overall decline in this service, but there were short-term fluctuations, which were more disruptive than a steady decline would have been. The total enrolment of merchant seamen dropped from a high of 46,911 in 1763 to 38,272 in 1765. In the following year, there was a temporary recovery with 44,599 enlisted, but by 1768 the number had again dropped, this time to 39,951.[63] The only branch of the Merchant Service which appears to have been steadily recovering after a severe depression in 1762 was the Greenland whale-fishing industry. But the numbers involved were relatively small.[64] There was, therefore, a large number of veterans of the Seven Years' War competing with unemployed seamen for a declining number of jobs in the 1760s. Many of the unemployed seamen congregated in the nation's major port, London.

Initially Parliament aided the demobilised veterans by applying an act of 22 George II which permitted them to work at trades without the customary apprenticeship requirements.[65] Some found work in waterfront or city trades by this means. While this

---

[60] *Shelburne Papers*, vol. 133, fol. 263.

[61] Ashton, *Economic Fluctuations in England, 1700–1800*, p. 187.

[62] 'Account of Seamen and Marines from 31 December 1765 to 31 December 1766', Add. MSS., 38340, fol. 11.

[63] Add. MSS., 38340, fol. 17.

[64] 'Account of the Number of Men Employed in the Greenland Whale Fishing' (dated 28 January 1768), Add. MSS., 38340, fol. 200.

[65] Ashton, *Economic Fluctuations in England, 1700–1800*, p. 152.

assisted the naval seamen to a degree, it was not a long-term solution to their problems, for it merely aggravated problems of overcrowding in such allied occupations as watermen and lighter-men. After October 1766 even this inadequate compensation was denied unemployed seamen who were war veterans, as a result of a precedent set in a case prosecuted by the Farriers' Company.[66] The seamen's strikes were therefore closely connected with other metropolitan industrial disputes of 1768, especially those which occurred among waterfront workers.

This labour surplus in the seafaring industry exerted pressure on the wages of those seamen fortunate enough to find work. Shipowners seized the opportunity of reducing their wage costs by various means. Some attempted to pay their men for the outward voyages and their homeward voyages separately. Thus their sailors received no pay for the time spent in foreign ports between trips.[67] This practice did not become general, but it does illustrate a general trend towards economies at the expense of seamen, which caused widespread unrest. More common was the circumvention of the apprenticeship system.[68] Ships carried more apprentices than was customary, as a means of reducing the complement of more expensive, trained seamen, carpenters and other trades. The length of apprenticeships was also reduced by many owners. Men who had not completed a full seven years' apprenticeship frequently served as able seamen or tradesmen at sea. One particular source of grievance was the practice of hiring three-year servants, which the owners justified on the grounds that they were 'a nursery for the Royal Navy'. The practice of carrying an excessive number of apprentices in allied occupations such as lightermen and watermen, because it denied them possible employment, was to the disadvantage of displaced merchant seamen too. The increased use of apprentices during periods of general depression was not solely due to the desire of owners to exploit cheap surplus labour to reduce costs. It also derived from the practice of apprenticing parish dependents, who increased during

[66] Humpherus, *History of Watermen and Lightermen of the River Thames*, p. 262.

[67] *Westminster Journal and London Political Miscellany*, 21 May 1768.

[68] See also Mantoux, *The Industrial Revolution in the Eighteenth Century*, p. 181, for reference to similar problems of framework knitters; George, *London Life*, pp. 180–1, 225–6 and *passim*; and Ashton, *An Economic History of England*, p. 224.

periods of economic difficulty. Certainly abuses of the apprenticeship system increased during the 1760s, and their correction
figured large in the demands of the seamen during their great
strike in May 1768.[69]

Like the coalheavers, the merchant seamen petitioned the
authorities to intervene in the spirit of the Elizabethan statute of
apprentices. But both the employers and the authorities were
already moving towards a loosening of labour-caste controls by
the 1760s. Although the latter were at times ambivalent in their
attitudes to changes in the economy, the exigencies of an expanding economy and the arguments of the exponents of *laissez-faire*
doctrines were taking effect. Caught in a period of change, the
seamen got the worst of both worlds, paternalistic and *laissez-
faire*. A legal judgement robbed them of the benefits of a more
liberal approach to the rehabilitation of veterans in shore trades,
while at the same time, in sea-going trades, a watering down of
the apprenticeship requirements resulted in greater competition
for jobs among the traditionally qualified ranks.

Again disturbances occurred in the provinces before they spread
to London. These various discontents manifested themselves in
riots and disorders in 1763.[70] Although comparatively limited in
scale, because they came at a time of general unrest among silk-
weavers and other industrial workers immediately after the war,
these disturbances caused the authorities concern. But it was not
until spring 1768 that really serious unrest showed itself among
seamen.

Mariners in north-eastern ports were the first to react against
worsening conditions. In early April 1768 crews demonstrated in
favour of higher wages to offset the rising prices of food, and
ships in the Tyne and Wear lay idle.[71] Like the coalheavers, the
seamen probably were encouraged to action by the success of the
northern pitmen.[72]

Here at Shields, Newcastle and Sunderland the seamen used
tactics which represent an interesting transitional stage between
the provincial poor's usual stress upon lowering prices by attacking middlemen and farmers, and the metropolitan poor's stress

[69] *Shelburne Papers*, vol. 133, fol. 331–74.
[70] *Calendar of Home Office Papers* (1760–65) I, no. 1168.
[71] *Annual Register*, XI (1768) 92.
[72] *Gentleman's Magazine*, XXXV (1765) 488.

on forcing employers to grant increased wages. The sailors prevented any ships leaving port, demonstrated, petitioned magistrates, attacked property and intimidated their employers in order to raise wages; while at the same time they attacked bakers and butchers to force down prices of 'necessaries'.[73] Like the hunger riots of 1766 these disturbances in the industrial centres of the north-east were stimulated by high prices and fear of famine.

While the tactics of the seamen were successful in eliciting promises of improved wages and conditions from the shipowners, the precise extent to which they lived up to their bargain is not apparent. Probably they reneged upon many of their undertakings as soon as the strike was over and the ships were at sea. They were not famed for honesty in their relations with their crews. In any case, the unrest now spread to London, partly as a result of events in the north, and partly as a result of general unrest among such Thames-side workers as coalheavers, lightermen and watermen. By early May 1768 all ships in the Thames, whether deep-sea or coastwise, were affected by the strike, which revealed the formidable organising skill of the seafarers.[74]

The strikers acted in the following manner. Committees of approximately forty men visited each ship in the river. They demanded to know the monthly wage rates, and to see the ship's articles. If the rates were below 40/- per month, they persuaded its crew to dismantle the rigging to prevent a hasty departure. The seamen then held assemblies in Stepney Fields to protest against conditions, and sent petitions to Parliament and the King begging for redress.[75] In this they acted in much the same fashion as the coalheavers, silk-weavers, tailors and the various other industrial groups who struck work and rioted in 1768. Although

[73] *Annual Register*, xi (1768) 92. *Domestic Entry Book*, vol. 142, fol. 44, 47–9, 50–1, 54, cited by Rudé, *Wilkes and Liberty*, p. 91.

[74] 'Memorials of Dialogues betwixt several seamen, a certain victualler, and a s—l master in the late riot', *Shelburne Papers*, vol. 133, fol. 363–74. These include a seaman's proclamation calling for a meeting of mates, carpenters and seamen at the Halfway House, Stepney Fields on Monday, 9 May 1768 'to consult proper measures for raising wages' and required ships' masters to meet with the seamen late the same day 'to settle and regulate the said seamen's wages'. The proclamation called on 'all watermen, lightermen, ballast men, ballast heavers, coalheavers etc. to leave their duty and not to go to work till our wages be settled'. It was signed on the left by 'no Wilkes, no King', and on the right by 'seamen'.

[75] *Annual Register*, xi (1768) 105–17

complaints at the high prices of provisions were frequent, no attacks were directed against middlemen of the provision trade. Representatives of the sailors met with the owners to present their grievances.[76] In the formulation of their demands the seamen sought the aid of victuallers, at least one of whom was prompt to inform the authorities of their intentions.

These demands were concerned primarily with increased wages and the ending of unfair competition for jobs. They asked for a monthly rate of 40/- in place of the current 32/-; a limitation of one apprentice per carpenter at sea and two per carpenter on land; an end to the practice of hiring three-year servants; enforcement of seven-year apprenticeship for seamen and other sea-going trades; limitation of two apprentices per master; and abandonment of the appointing of unqualified shipwrights as carpenters. While all these demands applied to the Merchant Service in general, one demand was relevant to the collier trade alone. They asked for 1/- per day waiting money for ships' crews delayed more than six days in the port of London, Shields, Sunderland or Blythe. This arose from the resentment of crews who were paid for a round trip irrespective of delays due to inclement weather or labour disputes of coalheavers.[77]

To what extent the seamen were successful in their demands in May 1768 is not readily apparent. Certainly they did not gain the 1/- per day waiting money. Their continued exasperation with the coalheavers resulted in an agreement with the owners to unload the coal themselves, with the bloody results already described. Certain shipowners were more ready to come to terms than others. Thus the Hudson's Bay Company agreed to pay its seamen the full 40/- per month, which it considered extortionate, because further delay might have meant that distant factories were not relieved that year.[78] Companies which dealt in perishable cargoes were also quick to settle.[79] Other owners agreed to a settlement with the men's representatives, but it apparently fell short of the men's demands because unrest continued and strikes

[76] Rockingham MSS., R1–1055, 13 May 1768.
[77] 'Memorials of Dialogues betwixt several seamen, a certain victualler, and a s—l master in the late riot', *Shelburne Papers*, vol. 133, fol. 363–74.
[78] Sir Mw Fetherstonhaugh to Newcastle, 19 May 1768, Add. MSS., 32990, fol. 107–8.
[79] *Public Advertiser*, 23 May 1768.

occurred as late as 24 August, when the seamen struck for 37/- per month.[80]

What success the seamen did enjoy was due to their economic and strategic importance to the country. Their strike had an immediate impact on the commercial life of London and threatened the national well-being. The possibility, however remote, of their sailing to France gave even greater cause for alarm, and accounts for the stationing of naval vessels to blockade the river.[81] Yet popular sympathy for their conditions aided the seamen's cause too. Most Englishmen held merchant and naval seamen in general affection and respect owing to their role in the late war. The sailors' disciplined restraint in their demonstrations, and their forbearance from joining with other political and social protesters gained them sympathy. On the other hand, the discipline and organisation evident in the punishment of criminals masquerading as seamen-protesters and the restoration of the spoils to their rightful owners must have frightened the authorities in a period when their peace forces were so limited.[82] The hint that there were 'forty thousand brave men at the northward' as well as 'fifteen or twenty-thousand Spitalfields weavers' who could 'join us at an hour's warning' suggested the interrelationship of the industrial disorders of 1768 and encouraged the authorities to deal cautiously with the rioters.[83]

Wisely the government avoided the provocation of using Guards to quell the strike. There was a natural antipathy between the soldiers and the seamen which did not exist between naval ratings and merchant seamen. Frequently soldiers and seamen were economic rivals. One paper reported a rumour that soldiers let their army pay go to officers and worked as coalheavers.[84] Legal documentation seems to corroborate this claim. One witness against two coalheavers was Francis Reynolds, 'a soldier in the guards who keeps a house in Shadwell'.[85] There was a tradition of moonlighting in the army. The Ministry's decision to

---

[80] Lloyd's Evening Post, 26–29 August 1768.

[81] 'Draught of a Letter from the Board of Admiralty to the Earl of Shelburne' (May 1768), Shelburne Papers, vol. 133, fol. 347–51.

[82] St. James's Chronicle, 12–14 May 1768.

[83] 'Memorials of Dialogues betwixt several seamen, a certain victualler, and a s—l master in the late riots', Shelburne Papers, vol. 133, fol. 363.

[84] Westminster Journal and London Political Miscellany, 30 May 1768.

[85] Treasury Solicitor's Papers, T.S. 11/818/2696. My italics.

withhold intervention by naval forces was based on entirely opposite considerations to those which determined the non-use of the soldiers. Here they feared that the ratings might side with the merchant seamen. The government therefore preferred to use its own agents to divert the seamen along lines of action most acceptable to it.

Almost certainly two at least of the several adjutants and one of the victuallers, who advised the seamen in drawing up their demands, represented the government's interests. The seamen gave the adjutants the right to negotiate a settlement with the owners at the King's Arms Tavern, Cornhill, on 13 May 1768. Lord Barrington, Minister-at-War, in a letter to Robert Wood, Under-Secretary to Lord Weymouth in the Northern Department, referred of one of these representatives as 'the bearer, Bell, who has been extremely useful in managing the seamen. . . .'[86] Edmund Burke named another of the leaders of the seamen's strike, Captain Fall, as an employee and pensioner of the Ministry, during a debate in the Commons in the following year.[87] Another critic of the government went even further when he accused the government of fomenting the seamen's strike for political purposes. There is no evidence of the government's provocation of such a dangerous strike, but in a century when mobs stirred up by the authorities were not unknown, their manipulation of riot leaders is entirely credible.

The efforts of these servants of the Ministry were threefold. First, they kept the government closely informed on the seamen's intentions; second, they discouraged the seamen from joining with other industrial protesters or political mobs; third, they persuaded the seamen to give them a *carte blanche* in negotiations and then agreed upon a settlement.

To sum up, the unrest of the seamen in 1768 resulted from the disruption of a period of trade realignment, which was exacerbated by high prices and the progressive abandonment of paternalistic protection. Seamen, like coalheavers, revealed their own uncertainties by their blend of traditional demands for protection and more progressive stress on improved wages and conditions. In this they were typical of the various industrial workers who

---

[86] *Calendar of Home Office Papers* (1766–9), no. 894, 23 May 1768.
[87] *Debates of the House of Commons* (1768–70), vol. 80, 7 April 1768.

rioted in the 1760s. This vacillation was paralleled by the attitudes and actions of the Ministry who were torn between the old
Elizabethan labour statutes and the more recent abstention from
interfering in trade. The employers sought to maximise their
profits by exploiting the surplus labour of their industry. Benefit-
ing from the growing disenchantment of the authorities with
outmoded protectionist policies, the shipowners eroded the
apprenticeship system and reduced the wages of their tradesmen.
In the process they alienated their seamen and stimulated the
strikes of 1768.

III

Of all the industrious poor of the metropolis who found their
conditions intolerable in 1768, the silk-weavers were most prone
to violent protest. Like seamen and coalheavers, they had a long
history of unrest, which stretched back at least as far as the early
years of the eighteenth century. Many of their resentments arose
from the innate problems of their trade.

Essentially the silk-weavers manufactured a luxury product,
the demand for which was quickly influenced by changes in
fashion or by economic vicissitudes. Thus even periods of court
mourning brought crisis to the silk trade. This occurred in the
1760s when chronic depression obliged the King to curtail court
mourning.[88] Inevitably reduced demand for consumer goods in
times of depression affected the luxury trades first. Such was the
case for example in early 1741 when, after a year of depression,
silk-weavers were reportedly starving by the thousands for lack
of work.[89]

Overseas markets for silk products were volatile too. English
exports to Europe received heavy competition from French and
Italian manufacturers, and after 1720 the British government
paid a bounty to encourage trade. This practice enabled the overseas silk trade to survive, but it did not solve the problem of an
overall decline of the metropolitan silk industry during the
eighteenth century.

This long-term decline was due to the inability of master silk-
weavers of London to match the lower costs of production of their

---

[88] *Westminster Journal and London Political Miscellany,* 25 January 1768.
[89] Ashton, *Economic Fluctuations,* p. 147.

competitors. One pamphleteer recommended the decentralisation of the industry by moving to Wales where labour costs were one-third those of Spitalfields, to meet French competition and reduce the problem of London's growth.[90] Another correspondent supported this argument when he wrote :

The true cause of these and the like disorders is that the price of wages in this town is too high to support a manufacture that does not require any very extraordinary skill and in which we are rivalled by other nations, the manufacture itself will by degrees probably remove to Glasgow but the quick decay of it here will in the meantime produce temporary convulsions.[91]

In the event, the convulsions were hardly temporary – they lasted into the next century – but the growing competition from the provinces proved correct.

Some producers accepted such advice, and competition for the metropolitan silk industry came not only from France and India by the mid-century, it also came from provincial centres like Glasgow, Coventry and Macclesfield, where labour and other costs were cheaper than in the capital. Disturbances amongst metropolitan silk-weavers must be set in the wider context of country-wide trade conditions. While tariff protection only partially solved the problem of foreign competition because of widespread smuggling, it did nothing to ameliorate the formidable rivalry of provincial weaving centres. Neither did mercantilist policies solve the chronic overcrowding of the silk-weaving trade.

The use of cheap apprentice-labour and the introduction of 'engine looms', each of which did the work of several hand-loom weavers, created bitter enmity between different categories of weavers, without finally solving the problems of high-cost production. Except in times of extraordinary activity, there was always an overwhelming surplus of labour. When trade was prosperous, the labour force expanded rapidly. This happened in wartime, when England diverted trade from her European competitors. In the years 1706–11, 1717–18, 1741–3 and

---

[90] 'Political Speculations on the Dearness of Provisions', anonymous pamphlet (London, 1767).
[91] *Shelburne Papers*, vol. 133, fol. 331–3 (n.d., between 1725 and 1749).

1756–62, the metropolitan industry expanded to satisfy a temporary wartime need.[92] With peace came renewed competition in overseas trade and a consequent reduced demand for labour. Clearly such fluctuations were more disturbing than a steady decline, for they over-expanded the labour force and created hopes among the industrious poor that were sooner or later dashed.

There were other reasons why there was a surplus of silk-weavers in most years of the century. Because silk-weaving demanded relatively little skill, it attracted recent arrivals in the capital. Irish immigrants in particular took up the trade, and settled in the east-end parishes known collectively as Spitalfields. Distress in Ireland or the English provinces increased the flow of immigrants to the capital. Thus in periods of widespread depression, more labour gravitated into the silk trade of London. This was the case following the agrarian hunger riots of 1766. Heavy immigration into London resulted in underemployment much of the time, and outright unemployment and starvation in times of economic slump. Such labour surpluses rendered the silk-weavers vulnerable to wage-cutting and other cost-saving devices.

In periods of depression and recession particularly, the master silk-weavers sought to reduce wage costs by various expedients. They lowered wage rates, encouraged the use of machinery and the employment of cheap labour. But the desire to meet cheap competition was not the only motivation for employers to rationalise their production at the expense of the journeymen. In an age when many commentators frequently noted the conspicuous consumption of the middling sort, master weavers sought to maximise their profits in order to compete socially with a variety of beneficiaries of economic change. This tendency was especially evident after the mid-century, and excited the resentment of the weavers.

Like the journeymen tailors who also rioted for better wages early in 1768, the response of the silk-weavers early in the century to worsening conditions had been to organise themselves into tightly-disciplined societies for the different branches of their trade.[93] These embryonic trade unions had concerned themselves with wages and conditions, as well as with the establishment of

92 Ashton, *Economic Fluctuations*, pp. 74–5.
93 *Shelburne Papers*, vol. 133, fol. 331–3, 19 November, n.d.

benevolent funds. They were fairly successful in frightening some concessions from master weavers, in contrast to the tailors, several of whom were prosecuted as committeemen of associations 'which raised a fund to support each other in such unlawful meetings, and who distinguished themselves by the name of Flints',[94] until an act of the sixth year of George III's reign made it a capital offence to engage in such combinations which hitherto had been tolerated despite their illegality. Forced underground by the hostility of the authorities, these societies met secretly at such hostelries as the Northumberland Arms, the White Hart, the Dolphin and the White Horse. They communicated in code, exchanged information on wage rates with weavers in Dublin and elsewhere in England, published cryptic messages in the press and maintained armed guards when in session. To support themselves these societies exacted a weekly or monthly toll of 2d, 3d or 4d per loom for master silk-weavers, and required all journeymen weavers to organise themselves into groups for the purpose of improving their living conditions. They issued formal receipts for moneys received in the name of, for example, the *Dreadnought*, which represented the broad silk-cutters, or the *Bold Defiance*, which represented the half-silk and half-worsted branch.

Committees of these various branches of the silk-weaving trade published books of pay rates. The Horsehair branch, for example, covered its expenses by demanding from master weavers 2d per book and 5d for each loom. In 1768, under these rates a weaver might have earned 12/- to 14/- per week had the masters not reversed their earlier decision to pay the wages demanded. The master weavers were encouraged in their defiance of the highly-organised weavers by the government's determination to crush the illegal combinations and the provision of military protection.

Certainly the close geographical concentration of workers in the east-end parishes of London aided the development of such societies dedicated to achieve improved conditions by whatever means necessary. The rejection of the Irish immigrants, who gravitated to the trade, by the lower orders of London's pre-industrial society ensured great social cohesion amongst silk-weavers, many of whom were Irish. The expatriated Whiteboys may have played as important a role in the silk-weavers' riots of

[94] *Annual Register*, VIII (1765) 79.

1768 as they did in the coalheavers' disputes of the same period. While oaths of membership for industrial groups were not peculiar to Irish terrorist societies in the eighteenth century, they may have spread as a result of Irish immigration.[95] In any case, there was plainly a significant exchange of information and co-operation between Dublin and London-based silk-weavers' societies.

If their organisation looked ahead to modern industrial organisation of labour, in their goals and tactics the silk-weavers displayed the same ambivalence as most of the industrious poor both of the town and country districts, who protested against their conditions in the second half of the eighteenth century. For the most part their goals were conservative. They wished for government regulation of their trade in the spirit of earlier, paternalistic times. To this end they petitioned the King and the Commons for relief. They paraded in time-hallowed fashion, frequently in very large numbers, to demand tariff protection against foreign competition, or to demonstrate their general dissatisfactions. They complained of the circumvention of apprenticeship regulations. But in their tactics of striking and attacking machinery or cutting threads in looms in order to impose an acceptable wage scale on their masters, silk-weavers revealed a more modern type of industrial radicalism more in keeping with their quasi-trade unionism.[96] These developments reached their peak in the decade between 1763 and 1773. They reflected the frustrations of the weavers at the progressive abandonment of wage and price-fixing of local authorities, severe increases in the price of food and post-war difficulties of the silk trade. It is now necessary to examine in

[95] George, 'The London Coalheavers', p. 236.

[96] Early in the eighteenth century, Spitalfields manufacturers were reportedly establishing combinations of 'greater extent and regularity' than ever before. They numbered 'several thousands' who were reduced 'to most exact discipline under their leaders, they plant sentinels in all the neighbourhood of Spitalfields and are ready to collect themselves on any alarm, they disguise themselves with crapes and are armed with cutlasses and other weapons, they write threatening letters in the form of humble petitions to the master manufacturers and they deter by threats those labourers from working at under price. . . .' Shelburne's correspondent went on to describe night raids to cut work and destroy looms by weavers who were 'learning the discipline of the regular troops'. Masters and others were too intimidated to give information (*Shelburne Papers*, vol. 133, fol. 331–3, n.d.). See also Mantoux, *The Industrial Revolution in the Eighteenth Century*, pp. 81–2; and J. L. and B. Hammond, *The Skilled Labourer, 1760–1832* (London: Longmans, 1933) pp. 205–10.

more detail the unique problems of the silk trade in the ten years which followed the Seven Years' War.

Like many of the trades of London, silk-weaving had enjoyed unusual prosperity during the war, because of the reduction of foreign competition and the new market opportunities. Trade had expanded rapidly. One writer noted the transference of haberdashers into silk manufacturing during this period of opportunity.[97] Much labour was attracted to the industry generally. Thus when a slump came at the end of the hostilities, and tariff protection was not resumed upon the signing of the peace treaty, the silk-weaving trade was badly hit by unemployment when the expectations of the industrious poor were high. Generally high food prices aggravated the problems of the weavers, and the next decade witnessed chronic unrest in the trade. Of these years, 1765 and 1768–9 were the most disturbed.

Immediately after the peace settlement, the silk trade suffered the loss of foreign and home markets, because of the revived French industry. Following very large demonstrations of both masters and journeymen of the silk-weaving trade in favour of the restoration of protection, serious rioting broke out in London. In this, only the intervention of the Guards saved the Duke of Bedford's residence from destruction. Frightened by the violence, Parliament now reimposed tariffs on foreign silks. Although it took several years for the exclusion of foreign silks to take effect because of stockpiling, this quietened the silk trade for a while. But continued depressed trade led to further conflict.[98] The pattern of action was much the same as earlier in the century, but the scale of action was much greater. Competition for scarce jobs exacerbated the rivalry between engine and hand-loom weavers.[99] Bloody encounters took place, and rioters attacked work in the looms and other property. Matters were not helped by a period

[97] Wages were high, and gauze and capuchin workers earned as much as 36/- in three days (*Gazetteer and New Daily Advertiser*, 30 January 1768).

[98] It was calculated in 1766 that there were three years' supplies of raw material stockpiled. Thus the embargo had little immediate effect other than psychologically to encourage the weavers.

[99] The *Public Advertiser* reported magistrates united in the opinion that a 'standard price of labour was wanting' between the labourers in all branches of the trade. Dorothy George claims that the dispute was over working below rates rather than over machinery displacing workers (George, *London Life*, p. 188).

of court mourning which, as earlier noted, the King shortened.

The most serious period of rioting in 1768–9 began with attempts in January 1768 by masters to reduce wage rates by 4*d* per yard, a gambit which was to stimulate similar reactions when attempted by shipowners, coal-undertakers, master tailors and other employers in the same year. Depressed conditions in the provincial silk-weaving centres made possible this action by master weavers. Rioters renewed their efforts to persuade Parliament to intervene in their favour, while at the same time they tried to coerce their employers to pay according to the agreed scales. Their actions encouraged similar responses from other groups in pre-industrial London. Matters reached a crisis point in late 1768 and early 1769. The wealthier master weavers armed their households in the face of the weavers' threats. Such actions stimulated warnings from the journeymen such as the following :

Cheavet;

> Your own house has been so armed of late that if you continue to hire men as you have done by giving them money and arms with persuasions and instructions how to use them whereby many innocent lives are in danger, the public committee held at this house – both by the assigns of master and men must be immediately obliged to make their application to the laws of the land to protect their lives as many evidence have declared, that you have sundry and false pretences encouraged them by offering of money and ordering liquors for them—whereby their senses were infused so as they did not know your wicked design and several of these persons have deposited the arms which they received to do what you required them, but they have declared they would not and have even marked the money or coin you gave them.
>
> From the public committee[100]

The government now, frightened by general industrial unrest and political riots, upon the advice of Sir John Fielding decided to suppress the 'unions' which were known to meet at various taverns of the metropolis. Among the most active magistrates in the move to crush the silk-weavers was Burford Camphire, a leading figure in the earlier coalheavers' disputes. The use of military forces led to casualties on both sides, but was successful in ending major outbreaks. Desultory attacks on individuals con-

[100] *Treasury Solicitor's Papers*, T.S. 11/818/2696.

tinued until the passage of the Spitalfield Act of 1773 restored paternalistic government control of the industry.[101]

This act represented a triumph for the weavers, for it realised their goal of restoring official regulation of the industry, and it compensated for the collapse of their more radical essays into collective bargaining in a developing climate of *laissez-faire*.[102] Although their standard of living continued to drop during most of the remaining years of the century, due to the inherent weaknesses of their trade, silk-weavers ceased to be the social threat they were in the 1760s.

From the viewpoint of the authorities, the act represented a reversal of a trend away from the principles of the old 'moral economy'. Like the earlier proclamation of the old statutes against middlemen during the food crisis of 1766, the government's action in 1773 represented a backward-looking step, but unlike the earlier regression the return to older principles in 1773 was not short-lived. By 1773 Parliament had already repealed the anti-middlemen statutes, revised the threshold prices for the export bounty on corn, and continued to favour the progressive abandonment of the assize of bread and wages. Thus Parliament's action in regulating the silk industry in 1773 was even more anomalous than earlier reversals of economic policies. It illustrates the confusion of the ruling orders in a period of rapid economic change and their ignorance of the developing economy. It solved neither the problems of the industry itself, nor those of the journeymen weavers, who continued to languish in poverty into the next century, with rare years of temporary partial recovery.

IV

The riots of the coalheavers, the seamen and the silk-weavers were typical of the pre-industrial disorders in the metropolis of the 1760s in their form and direction. The labouring poor at this time began to adopt trade union tactics, and sought higher wages

---

[101] S. and B. Webb, *The History of Trade Unionism*, pp. 54–5.

[102] The government took the advice of Sir John Fielding to give clear authority to the magistrates to set wage rates ('Cutters of Looms in Spitalfields, 28 September–18 December 1769', *State Papers*, SP 37/7, fol. 85–108).

in face of rising costs rather than a reduction of prices. Yet they exhibited a curious mixture of old and new tactics and goals. Strikes for better conditions and higher wages were always combined with demands for government intervention to preserve privileged positions or to control the employers. Only in their extent and violence did the riots of these three groups of London workers differ from those of the tailors, shoemakers, coopers and watermen. The prolonged nature of the protests of coalheavers, seamen and silk-weavers probably indicate their tighter organisation and the greater influence of disciplined, radical elements such as veterans or Irish terrorists. Their close proximity in time to the Wilkite disorders rendered them most effective, for the forces of order were already greatly extended. Sympathy strikes did not occur despite the government's fear that seamen and silk-weavers might combine. But there is little doubt that disorders were contagious. Certain provincial disputes appear to have set off chain-reactions in the trades of the metropolis. The violent industrial dispute between miners and coalowners in the north-east in 1765 was followed by a strike of seamen on colliers, which spread to all seamen in London, encouraged the coalheavers to riot, and ultimately affected all trades on the London waterfront. The role of the seamen as a vehicle for conveying radical ideas and methods of action is an important one here. The contagion of violence seems to have spread from the provinces to London's outward dockland and from there through the Tower Hamlets and eventually into Spitalfields in the east end of London. This radiating action, however, was not the only pattern of disorder. Riots among other trades in the metropolis preceded this movement, and the causes of discontent were deep seated in other trades, which accounts for their sensitivity to the climate of unrest which spread from the waterfront. The mixture of old and new tactics indicates that these disorders were a transitional stage between the street protests touched off by sudden price rises on the one hand and organised industrial action which emerged in the next century on the other. While food protests in the traditional form of 'taxation populaire' continued into the next century in rural regions, already in the 1760s the metropolitan industrial protests had more in common with modern industrial tactics than the traditional petitions and parades demanding government intervention to protect the workers. Only the silk-weavers' subsequent history belies this contention. They alone maintained their traditional goals of paternalistic protection from the government. Their very success however resulted in the declining living standards of their overpopulated trade. Had they failed to get government protection, it is probable that, subjected to the full competition of the market, many of the poor would have ceased to gravitate to a trade in which government protection kept members from outright destitution, but only just.

These riots also reveal the confusion of the government when faced with serious industrial unrest, which contributed to the disorders. Unable to assess accurately the causes of discontent, and ignorant of the operation of the English economy, they first fell back on outmoded methods of dealing with economic distress, and then vacillated between encouraging rioters to solve their own problems and punitive measures of suppression. Most frequently they relied upon a few able magistrates in their pay to maintain order, in the last resort by the use of troops. They also used paid informers to manipulate the rioters where possible. Even when, as in the case of the coalheavers, the exploiters of the poor and stimulators of protests were recognised, the government chose to tolerate such groups as the coal undertakers when they appeared to be able to maintain a semblance of order among the 'depraved' residents of the Tower Hamlets. Presumably on the principle of setting a thief to catch a thief they appointed successful manipulators like Green to be magistrates over the rebellious poor.

The industrial riots of 1768–9 had much in common with the earlier, rural hunger riots. The latter represented a less sophisticated form of social protest than the former. In both forms of social protest, underlying tensions gave form and direction to the disturbances. The provincial hunger riots formed the immediate background to the industrial riots, although each of the latter had a unique history of distress. With the available evidence one may only conclude that there was considerable interaction between provincial and metropolitan industrial disputes by 1768. But the labouring poor had little sense of cohesion. Their loyalty lay with their trade rather than with any 'class'. Any co-operation was entirely on an *ad hoc* basis. As usual at this time, the authorities' apprehensions about a great conspiracy were not justified.

Probably in their effect on the authorities, and the ruling

classes generally, the riots of the 1760s were most significant. The commitment to less government intervention in the economy was one of the results of the experiences of the 1760s. Repeal of the anti-middlemen legislation, modification of the Corn Laws and a progressive abandonment of the old 'moral economy' were evident examples of this. Perhaps the closing of ranks among the landowners, farmers and industrialists against the threat of future insurrection was of greatest significance.

Conclusions

This study has examined both the immediate and underlying causes of social protest in the first decade of George III's reign. The evidence points to sudden fluctuations in the prices of food, grain movements in times of apprehended famine or attempts by employers to reduce wages or employment opportunities, as factors which precipitated many of the provincial and metropolitan riots of the 1760s. Because comparable disorders did not occur at other times in the century when deprivation among the poor appears to have been greater, the expectations of both the rioters and other important interests, such as the landowners and industrialists, were of great importance. The relativity of expectations was manifested in the significant role of the veterans of the Seven Years' War in the disorders, and in the equivocal attitude of the local and national authorities towards the initial outbreaks. It has been shown that social tensions, created particularly by agrarian and industrial developments and intensified by war and the progressive abandonment of the paternalistic regulation of the economy which had reached its peak in Tudor and Stuart times, underlay such expectations. One major determinant of the direction that hunger riots took was the popular prejudice against the monopolistic practices of farmers and middlemen of the provisions trade. Fostered by newspapers and periodicals, as well as government policies during periods of dearth, the growing antipathy towards these interests provided a climate of opinion in which popular scapegoats could be presented. By seizing the opportunity, the ruling orders of the countryside were able to divert the dispossessed against large farmers and middlemen, and to avoid dangerous social isolation during the hunger riots of 1766.

But industrial and commercial interests in the metropolis in 1768 also benefited from the ambivalence of the magistrates and the government towards the reassertion of the statutes for the protection of both the consumers and the workers. Generally local and national authorities abandoned their efforts to return to the certainties of an earlier age, after their initial attempts to enforce the requirements of the old 'moral economy'. While some saw that the economy had outgrown the ability of the bureaucracy to regulate it and that the restoration of partial control only exacerbated existing problems, others of the ruling orders vacillated between paternalism and non-intervention. The inconsistencies of a period of transition in economic thought and practice are apparent in the uneven abandonment across the country of the assizes of bread and wages, the inadequate legislation to protect the coalheavers from the coal-undertakers and their later surrender to the unregulated control of these middlemen; and the relaxation of the Statute of Apprentices and the paternalistic Spitalfields Act. Yet the trend after the 1760s was away from paternalism to a freer economy. In times of crisis, individual magistrates and most of the industrious poor looked for a return to the old paternal regulations, but by and large the ruling orders found in the experience of the 1760s a confirmation of the theories of the new school of political economy articulated by Adam Smith. In legislation this trend was typified by the act which in the early 1770s repealed the anti-middlemen statutes.

Another lesson which the rural gentry in particular drew from the riots of the 1760s was the danger of divisions between various interests in the face of social insurrection, and of delayed repression of disorders. The tactic of separating the larger tenant farmers from the labourers who protested against the high cost of provisions in 1766 was successful; but the seriousness of the disturbances that developed with the sanction of rural magistrates taught both the gentry and the tenant farmers that in future riots their interests lay in co-operation. In fact there was a conservative reaction in rural England of the 1760s which found its natural outlet in the next century when rural magistrates took the lead in repressing disorders and urged their urban counterparts to snuff out rebellion before it grew to significant proportions. The events of the American Revolution, the Gordon riots and the radical threat during the French Revolution confirmed the gentry in their reaction. At the national government level, reaction expressed itself first in a hardening of attitudes to American revolutionaries and Irish rebels, and later in a rejection of even mild political reforms. Certainly the ruling orders were quicker to perceive the possibilities of popular movements than were the poor themselves. Government leaders were right in believing that veterans presented a serious threat when they turned to organising riots. The experiences of the 1790s and of the first two decades of the nineteenth century confirmed this. Yet the rural hunger riots did have one positive effect on English constitutional development. The increasingly frequent practice of constituencies 'instructing' their Members of Parliament was a direct result of local dissatisfaction with national policies which increased the distress of the poor and thereby threatened order.

The danger of reading too much into the popular protests of the 1760s in the light of subsequent events has been noted by others. Modern industrial action lay a long way ahead and agricultural unionism had yet to endure a hundred years of privation, riot and ferocious suppression. But the pattern of events in the 1760s represents a fascinating mixture of old and new. It is a decade of social transition which will continue to attract the scholar.

Select Bibliography

PRIMARY SOURCES

Manuscript Sources

British Museum, Additional MSS. 32732.
—— Add. MSS. 32875.
—— Add. MSS. 32977.
—— Add. MSS. 32990. [*Newcastle Papers.*]
—— Add. MSS. 35607. [Hardwicke MSS.]
—— Add. MSS. 38205. [Liverpool MSS.]
—— Add. MSS. 38340. [*Liverpool Papers.*]
—— Egerton MSS. 215. [*Cavendish Debates.*]
—— 'The Case of Mr. Francis Reynolds'.
—— 'The Coalheavers' Case', 1764.
Bury St Edmunds, West Suffolk County Record Office. Papers of Augustus Fitzroy, 3rd Duke of Grafton: Papers and Memoranda on Domestic Affairs, 1762–9. [*Grafton Papers.*]
William L. Clement Library, Ann Arbor, Michigan. *Grenville Papers*, vols I–IV.
—— Ligonier MSS. *Letter Book*, 1759–60.
—— *Shelburne Papers*, vols 111, 124, 130, 132–3, 135, 168.
—— Thomas Townshend, 1st Viscount Sydney, 1733–1800. *Sydney Papers*, 15 vols (1685–1829).
House of Lords Record Office, Westminster. *Papers of the Committee on the High Prices of Provisions*, March 1765. Main Papers.
Ipswich and East Suffolk Record Office, Ipswich. Albermarle MSS. 4050.
—— *Barrington Papers*.
—— *Manuscript Letter Book of Viscount Barrington*.
Public Record Office, London. *Admiralty Entry Book*. 1766–84.

—— *Chatham Papers.* PRO 30/8/33; PRO 30/8/56; PRO 30/8/66.

——*Marching Orders of the Army.* WO5–54; WO5–55; WO5–56.

—— *State Papers Domestic.* PRO/SP 37/4; PRO/SP 37/6.

—— *State Papers. Domestic Entry Book,* vols 25, 141–2.

—— *Treasury Solicitor's Papers.* T.S. 11/443/1408; T.S. 11/818/2696; T.S. 11/290/3213; T.S. 11/995/3707; T.S. 11/1116/5728; T.S. 11/5956/Bx1128.

Sheffield Central Library. *Burke Papers,* bk 86.

—— *Rockingham Papers.* 1–706; 1–737; 1–1052; 1–1054; 1–1055; 1–1058(a); 1–1058(b); 1–1062; 1–1063; 1–1064.

—— *Wentworth Woodhouse Muniments.* [Fitzwilliam MSS.] 1–105; 1–107; 1–166.

Local Government Records

Cambridge Record Office, Cambridge. *Cambridge Sessions Books,* 1766–8.

Derbyshire Record Office, Derby. *Derby Quarter Sessions Order Book,* 1766.

Gloucestershire Record Office, Gloucester. 'Draft of Petition of Gloucester Justices for a Standard Measure of Corn, 1709.' D214/b10/4.

—— *Quarter Sessions Order Book,* 1766–80.

Middlesex Record Office, Westminster. *Middlesex Poll Book,* 1768–9.

—— *Order Book.* Middlesex General Quarter and General Sessions of the Peace. 3–14 George III, no. 8.

——*Process Register Book of Indictments.*

—— *Session Papers,* 19 May 1768, 'Petition for Redress by the Inhabitants of St. Paul's, Shadwell and Adjacent Places in the County of Middlesex'.

—— *Session Papers,* September 1768, 'Representation of Justices for Middlesex in General Session Assembled at Hick's Hall, Thursday, September 8, 1768'.

Norwich City Record Office, Norwich. Box of *Depositions and Case Papers.*

—— *Norfolk Quarter Sessions Order Book,* 1752–74.

—— *Norwich Quarter Sessions Order Book,* 1755–75.

Printed Correspondence, Memoirs and Journals

Calendar of Home Office Papers of the Reign of George III, ed. Joseph Redington, vol. I (1760–5) and vol. II (1766–9) (London: H.M. Stationery Office, 1878–9).

Correspondence of John, Fourth Duke of Bedford, vol. III (London: Longman, 1846).

Dobrée, Bonamy, ed. *Letters of George III* (London: Cassell, 1968).

Fortescue, Sir John William, ed., *George III, King of Great Britain, 1738–1820: The Correspondence of King George the Third*, vol. I (1760–67). 1st ed. New impression (London: Cassell, 1967).

The Grenville Papers: Being the Correspondence of Richard Grenville Earl Temple, K.G. and the Right Hon. George Grenville, their Friends and Contemporaries, ed. William James Smith, vols III–IV (London: John Murray, 1853).

House of Commons Journals, 1765–9.

House of Lords Journals, 1765–9.

Macpherson, David, *Annals of Commerce, Manufactures, Fisheries and Navigation . . . Containing the Commercial Transactions of the British Empire and Other Countries, from the Earliest Accounts to . . . January 1801; and Comprehending the Most Valuable Part of . . . Mr. Anderson's History of Commerce, viz. from the Year 1492 to the End of the Reign of George II, etc.*, 4 vols (London, 1805).

Parliamentary History of England, 1765–1771, XVI (London: Longmans, 1813).

Walpole, Horace, *Letters*. Edited by Paget Toynbee, 19 vols (Oxford: Clarendon Press, 1903–25).

—— *Memoirs of the Reign of King George the Third*, re-edited by G. F. Russell-Barker, vol. II (London: Lawrence & Bullen, 1894).

The Journal of the Reverend John Wesley, A.M., ed. Nehemiah Curnock, Standard Edition, vol. V, 2nd ed. (London: Charles H. Kelly, 1781).

Young, Arthur, *A Six Weeks' Tour through the Southern Counties of England and Wales* (London: W. Nicoll, 1768).

—— *The Farmer's Tour through the East of England*, 4 vols (London: W. Strahan, 1771).

Periodicals

Annual Register, IV–XII (1756–69).
Gazetteer and New Daily Advertiser, 1766–9.
Gentleman's Magazine, V (1735); VIII–IX (1738–9); XXVI–XXXIX (1756–69).
Gloucester Journal, 1757.
Lloyd's Evening Post, 1766–9.
London Chronicle, 1766–9.
Monthly Review, XXXVII (1767).
Norwich Mercury, 1757.
Public Advertiser, 1766–9.
St. James's Chronicle, 1766–9.
Universal Magazine, XLIII (1768).
Westminster Journal and London Political Miscellany, 1768.

Contemporary Pamphlets

A Letter to the House of Commons in which is Set Forth the Nature of Certain Abuses Relative to the Articles of Provisions, both with Respect to Men and Horses: together with their Remedies (London: J. Almon, 1765).
A Letter to the Right Honourable Lord North, 1772.
An Attempt to Discover the Causes of the Dearness of Provisions and the High Price of Labour in England with some Hints for Remedying these Evils (London: J. Almon, 1767).
Burke, Edmund, *Thoughts on the Cause of the Present Discontents, together with Observations on a Late Publication Intituled 'The Present State of the Nation'* (London: G. Routledge & Sons Ltd, 1913).
Colquhoun, Patrick, *Treatise on the Commerce and the Police of the River Thames: Containing an Historical View of the Trade of the Port of London; and Suggesting Means for Preventing the Depredations thereon, by a Legislative System of River Police, etc.* (London: Joseph Mawman, 1800).
Considerations on the Effects which Bounties Granted on Exported Corn, Malt, and Flour, 1768.
Considerations on the Exportation of Corn wherein the Principal Arguments Produced in Favour of the Bounty are Answered

and the Inferences Commonly Drawn from the Eton Register are Disproved (London, 1766).

Hanway, Jonas, Letters on the Importance of the Rising Generation of the Labouring Part of our Fellow-Subjects, vol. I (London, 1767).

Observations and Examples to Assist Magistrates in Setting the Assize of Bread Made of Wheat under the Statute of the 31st George II (London, 1759).

'Political Speculations on the Dearness of Provisions' (London, 1767).

Smith, Charles, Three Tracts on the Corn Trade and the Corn Laws, 1766.

Two Letters on the Flour Trade, and the Dearness of Corn: By a Person in Business (London, November 1766).

SECONDARY BOOKS

Books

Anonymous, The History of the City of Norwich: From Earliest Records to the Present Time (Norwich: W. Allen, 1869).

Ashton, Thomas Southcliffe, Economic Fluctuations in England, 1700–1800 (Oxford: Clarendon Press, 1959).

—— An Economic History of England: The Eighteenth Century (London: Methuen, 1955).

—— and Sykes, Joseph, The Coal Industry of the Eighteenth Century (Manchester: Manchester University Press, 1964).

Baker, T. H., ed., Records of the Seasons, Prices of Agricultural Produce . . . in the British Isles (London: Simpkin, Marshall, 1911).

Barnes, Donald Grove, A History of the English Corn Laws from 1660–1846 (New York: Augustus M. Kelley, 1961).

Barrington, Shute, The Political Life of William Wildman, Viscount Barrington, 1717–1793 (London: W. Bulmer, 1814).

Beloff, Max, Public Order and Popular Disturbances, 1660–1714 (London: Cass, 1963).

Besant, Walter, London in the Eighteenth Century (London: Adam & Charles Black, 1902).

Beveridge, William [Lord Beveridge], Prices and Wages in England from the Twelfth to the Nineteenth Century, vol. I, 2nd ed. (London: Frank Cass, 1965).

Bischoff, James, *A Comprehensive History of the Woollen and Worsted Manufacturers*, vols I–II (London: Smith Elder, 1842). 1st ed., new impression (London: Cass, 1968).

Bleackley, Horace William, *Life of John Wilkes* (London: John Lane, The Bodley Head, 1917).

Briggs, Asa, 'The Age of Improvement', *A History of England*, ed. W. N. Medlicott (London: Longmans, 1959).

—— and Saville, John, eds, *Essays in Labour History*. Papermac (London: Macmillan, 1967).

Brooke, John, *The Chatham Administration, 1766–1768* (London: Macmillan, 1956).

Chambers, J. D., and Mingay, G. E., *The Agricultural Revolution, 1750–1880* (London: Batsford, 1966).

Christie, Ian R., *Crisis of Empire, Great Britain and the American Colonies, 1754–1783*, Foundations of Modern History, ed. A. Goodwin (London: Arnold, 1966).

Clapham, Sir John Harold, *The Woollen and Worsted Industries . . . with Diagrams and Illustrations* (London: Methuen, 1907).

Clark, Sir George Norman, *Guide to English Commercial Statistics—1696–1782* (London: Royal Historical Society, 1938).

Clifford, James, ed., *Man Versus Society in Eighteenth Century Britain: Six Points of View* (Cambridge: Cambridge University Press, 1968).

Cole, G. D. H., *Short History of the British Working Class Movement, 1787–1947*, revised and enlarged (Aberdeen: George Allen & Unwin, 1948).

—— and Postgate, Raymond, *The Common People, 1746–1946*, 4th ed. (London: Methuen, 1949).

Copeland, Thomas W., ed., *The Correspondence of Edmund Burke*: vol. I, April 1744–June 1768, ed. T. W. Copeland (Cambridge: Cambridge University Press, 1958); vol. II, July 1768–June 1774, ed. Lucy S. Sutherland (Cambridge: Cambridge University Press, 1960).

Darvall, Frank Ongley, *Popular Disturbances and Public Order in Regency England* (London: Oxford University Press, 1934).

Deane, Phyllis, and Cole, William Alan, *British Economic Growth, 1688–1959, Trends and Structure*, 2nd ed. (Cambridge: Cambridge University Press, 1967).

De L. Mann, Julia, ed., *Documents Illustrating the Wiltshire*

Textile Trades in the Eighteenth Century, vol. xix (Devizes : Wiltshire Archaeological and National History Society, Records Branch, 1964).

Fay, Charles Ryle, *The Corn Laws and Social England* (Cambridge : Cambridge University Press, 1932).

Fletcher, Reginald J., B.D., ed., *Pension Book of Grays Inn, Records of the Honourable Society, 1669–1880* (London : Chiswick Press, 1910).

George, M. Dorothy, *London Life in the Eighteenth Century* (London : Kegan Paul, Trench, Trubner & Co. Ltd, 1925; New York : Harper Torchbooks, 1965).

Gilboy, Elizabeth Waterman, *Wages in Eighteenth Century England*, vol. xlv of *Harvard Economic Studies* (Cambridge, Mass. : Harvard University Press, 1934).

Gras, Norman Scott Brien, *The Evolution of the English Corn Market from the Twelfth to the Eighteenth Century* (Cambridge, Mass. : Harvard University Press, 1915).

Habakkuk, H. J., 'The Land Market in the Eighteenth Century', *Britain and the Netherlands*, ed. G. S. Bromley and E. H. Kossman (London : Chatto & Windus, 1960).

Hammond, J. L., and Hammond, Barbara, *The Village Labourer, 1760–1832*, 4th ed. (London : Longmans, 1932).

—— *The Skilled Labourer, 1760–1832* (London : Longmans, 1933).

—— *The Town Labourer*, 2nd ed. (London : Longmans, 1966).

Harris, J. H., ed., *A Series of Letters of the First Earl of Malmesbury, His Family and Friends from 1745 to 1820*, vol. i (London : R. Bentley, 1870).

Heaton, Herbert, *The Yorkshire Woollen and Worsted Industries: From the Earliest Times up to the Industrial Revolution*, vol. x of *Oxford Historical and Literary Studies* (London : Oxford University Press, 1965).

Hecht, J. Jean, *The Domestic Servant Class in Eighteenth Century England* (London : Routledge & Paul, 1956).

Highmore, Anthony, *The History of the Honourable Artillery Company of the City of London* (London, 1804).

Hill, Christopher, *Reformation to Industrial Revolution; The Making of Modern English Society, 1530–1780*, vol. i (New York : Random House, 1967).

Horn, D. B., and Ransome, Mary, eds, *English Historical Docu-*

ments, 1714–1783, vol. x (London: Eyre & Spottiswoode, 1957).

Hoskins, W. G., 'The Population of an English Village 1086–1801—A Study of Wigston Magna', *Provincial England* (New York: Macmillan, 1963).

Humpherus, Henry, *History of the Origin and Progress of the Company of Watermen and Lightermen of the River Thames*, vol. ii (London, 1887).

Jesse, John Heneage, *Memoirs of the Life and Reign of George III* (London: R. Bentley, 1852).

John, A. H., 'The Course of Agricultural Change, 1660–1760', *Studies in the Industrial Revolution*, ed. L. S. Pressnell (London: Athlone Press, 1960).

Jones, Erich L., *Seasons and Prices; The Role of the Weather in English Agricultural History* (London: Allen & Unwin, 1964).

Kerr, Barbara, *Bound to the Soil, A Social History of Dorset, 1750–1918* (London: John Baker, 1968).

Lecky, William Edward Hartpole, *History of England in the Eighteenth Century*, vol. iii (London: Longmans, Green, 1878).

Levebvre, Georges, *La Grande Peur* (Paris: Société d'Édition d'Enseignement Supérieur, 1956).

Lipson, Ephraim, *The History of the Woollen and Worsted Industries, Histories of English Industries*, ed. E. Lipson (London: A. & C. Black, 1921).

Maccoby, Simon, *English Radicalism, 1762–1785, The Origins*, vol. i of *The English Radical Tradition, 1763–1914*, ed. Alan Bullock and F. W. Deakin (London: Nicholas Kaye, 1955).

Mantoux, Paul, *The Industrial Revolution in the Eighteenth Century*, rev. ed. (London: Methuen, 1966).

Marshall, Dorothy, *The English Poor in the Eighteenth Century: A Study in Social and Administrative History* (New York: Augustus M. Kelley, 1926).

—— *Eighteenth Century England, A History of England*, ed. W. N. Medlicott (London: Longmans, 1962).

Mather, Frederick Clare, *Public Order in the Age of the Chartists* (Manchester: Manchester University Press, 1959).

Mingay, G. E., *English Landed Society in the Eighteenth Century* (London: Routledge & Kegan Paul, 1963).

—— *Enclosure and the Small Farmer in the Age of the Indus-trial Revolution*, Studies in Economic History, Papermac (London : Macmillan, 1968).

Namier, Sir Lewis, and Brooke, John, *The History of Parliament, The House of Commons, 1754–1790*, vol. III (London : H.M. Stationery Office, 1964).

Norris, John M., *Shelburne and Reform* (London : Macmillan, 1963).

Peacock, A. J., *Bread or Blood: A Study of the Agrarian Riots in East Anglia in 1816* (London : Victor Gollancz, 1965).

Postgate, Raymond W., *That Devil Wilkes* (New York : Van-guard Press, 1929).

Poynter, J. R., *Society and Pauperism, English Ideas on Poor Relief, 1795–1834* (London : Routledge & Kegan Paul, 1969).

Prothero, Rowland Edmund, Baron Ernle, *English Farming Past and Present* (London : Longmans, 1912).

Rudé, George, *The Crowd in History: A Study of Popular Dis-turbances in France and England, 1730–1848* (New York : John Wiley & Sons, 1964).

—— *Wilkes and Liberty* (Oxford : Oxford University Press, 1962).

Schumpeter, Elizabeth Boody, *English Overseas Trade Statistics, 1697–1808* (Oxford : Clarendon Press, 1960).

Thompson, E. P., *The Making of the English Working Class* (London : Victor Gollancz, 1963).

Tooke, Thomas, and Newmarch, William, *A History of Prices and of the State of the Circulation, from 1793 to the Present Time*, 6 vols (London, 1838–57).

Townsend, James, ed., *News of a Country Town: Being Extracts from 'Jackson's Oxford Journal' Relating to Abingdon, 1753–1835 A.D.* (London : Humphrey Milford, 1914).

Veitch, George Stead, *Genesis of Parliamentary Reform*, with an Introduction by I. R. Christie (London : Constable, 1964).

Victoria County History, Gloucestershire, ed. William Page, F.S.A., vol. II (London : Constable, 1907).

Victoria County History, Leicestershire, ed. W. G. Hoskins, vol. II (Oxford : Oxford University Press, 1954).

Victoria County History, Wiltshire, ed. Elizabeth Crettall, vol. IV (Oxford: Oxford University Press, 1959).

Wearmouth, Robert Featherstone, *Methodism and the Common*

People of the Eighteenth Century (London: Epworth Press, 1945).

Webb, Sidney, and Webb, Beatrice, *The Parish and the County*, vol. I of *English Local Government* (Hampden: Archon Books, 1963).

——*The English Poor Law. Part I: The Old Poor Law* (Hampden: Archon Books, 1963).

—— *The History of Trade Unionism*, 1894: Reprints of Economic Classics (New York: Augustus M. Kelley, 1965).

Westerfield, Ray Bert, *Middlemen in English Business, Particularly between 1660 and 1760*, vol. XIX of *Transactions of the Connecticut Academy of Arts and Sciences, 1866* (New Haven, Conn.: Yale University Press, 1915).

Western, John R., *The English Militia in the Eighteenth Century*, Studies in Political History, ed. Michael Hurst (London: Routledge & Kegan Paul, 1965).

Williams, Gwyn A., *Artisans and Sans-Culottes, Popular Movements in France and Britain during the French Revolution*, Foundations of Modern History (London: Arnold, 1968).

Wilson, Charles, *England's Apprenticeship, 1603–1763*, Social and Economic History of England, ed. Asa Briggs (London: Longmans, 1965).

Journal Articles

Ashton, T. S., 'Changes in Standards of Comfort in the Eighteenth Century', *Proceedings of the British Academy*, XLI (1955) 171–87.

—— 'The Coalminers of the Eighteenth Century', *Economic History*, Supplement to *Economic Journal*, I (1926–9) 307–34.

Beresford, M. W., 'The Common Informer, the Penal Statutes and Economic Regulations', *Economic History Review*, 2nd ser., x (1957–8) 221–37.

Blackman, Janet, 'The Food Supply of an Industrial Town', *Business History*, v (1962) 83–97.

Brown, E. P. H., and Hopkins, S. V., 'Seven Centuries of Building Wages (in Southern England)', *Economica*, XXII (1955) 195–206.

—— 'Seven Centuries of the Prices of Consumables, Compared with Builders' Wage Rates', ibid., XXIII (1956) 296–314.

—— 'Seven Centuries of Wages and Prices; Some Earlier Estimates', ibid., xxviii (1961) 30–6.

Campbell, Mildred, 'English Emigration on the Eve of the American Revolution', *American Historical Review*, lxi, no. 1 (1955) 1–20.

Chaloner, W. H., 'Recent Work on Enclosure, the Open Fields and Related Topics', *Agricultural History Review*, ii (1954).

Chambers, J. D., 'Enclosure and Labour Supply in the Industrial Revolution', *Economic History Review*, 2nd ser., v, no. 3 (1953) 319–43.

Coats, A. W., 'Economic Thought and the Poor Law in the Eighteenth Century', ibid., xiii (1960) 39–51.

Cole, W. A., 'Trends in Eighteenth Century Smuggling', ibid., x (1958) 395–410.

Davis, Ralph, 'English Foreign Trade, 1700–1774', ibid., xv (1962–3) 285–303.

—— 'Seamen's Sixpence; An Index of Commercial Activity 1697–1828', *Economica*, new ser., xxiii (1956) 328.

Fay, C. R., 'Significance of the Corn Laws in English History', *Economic History Review*, 1st ser., i (1927–8) 314.

Fisher, F. J., 'The Development of the London Food Market 1540–1640', ibid., v–vi (1935) 46–64.

George, M. Dorothy, 'The Early History of Registry Offices', *Economic History, Supplement to Economic Journal*, i, no. 4 (1926–9) 570–90.

—— 'The London Coalheavers: Attempts to Regulate Waterside Labour in the Eighteenth and Nineteenth Centuries', ibid., 229–48.

Granger, C. W. J., and Elliott, C. M., 'A Fresh Look at Wheat Prices and Markets in the Eighteenth Century', *Economic History Review*, 2nd ser., xx, no. 2 (1967) 257–65.

Habakkuk, H. J., 'English Landownership, 1680–1740', ibid., 1st ser., x, no. 1 (1940) 2–17.

—— 'English Population in the Eighteenth Century', ibid., 2nd ser., vi, no. 2 (1953) 117–33.

—— 'Essays in Bibliography and Criticism: The Eighteenth Century', ibid., viii (1956) 437–8.

John, A. H., 'Agricultural Productivity and Economic Growth in England, 1700–1760', *The Journal of Economic History*, xxv (1965) 19–34.

—— 'Aspects of English Economic Growth in the First Half of the Eighteenth Century', *Economica*, xxviii (1961) 181–7.

Jones, E. L., 'Agriculture and Economic Growth in England, 1660–1750 : Agricultural Change', *Journal of Economic History*, xxv (1965) 1–18.

Lloyd-Pritchard, M. F., 'The Decline of Norwich', *Economic History Review*, 2nd ser., iii (1950–1) 371–7.

Macdonald, Forest, 'The Relation of the French Peasant Veteran of the American Revolution to the Fall of Feudalism in France', *Agricultural History*, xxv (1951) 151–61.

Maier, Pauline, 'John Wilkes and American Disillusionment with Britain', *William and Mary Quarterly*, 3rd ser., xx (1963) 373–95.

Mathias, Peter, 'Agriculture and the Brewing and Distilling Industries in the Eighteenth Century', *Economic History Review*, 2nd ser., v, no. 2 (1952) 249–57.

Mingay, G. E., 'Size of Farms in the Eighteenth Century', ibid., xiv (1961–2) 469–88.

—— 'The Agricultural Depression, 1730–50', ibid., viii (1956) 323–38.

Rose, R. B., 'Eighteenth Century Price Riots and Public Policy in England', *International Review of Social History*, vi, no. 2 (1961) 277–92.

Rudé, George F. E., 'The London "Mob" of the Eighteenth Century', *Historical Journal*, ii, no. 1 (1959) 1–18.

—— 'The Study of Eighteenth Century Popular Movements', *Amateur Historian*, iv, no. 6 (1959–60) 235–41.

Skeel, Caroline, 'The Cattle Trade between Wales and England from the Fifteenth to the Nineteenth Century', *Transactions of the Royal Historical Society*, iv (1926) 135–58.

Southwood, Ken, 'Riot and Revolt: Sociological Theories of Political Violence', *Peace Research Reviews*, i, no. 3 (1967).

Stone, Lawrence, 'The Ninnyversity?', *New York Review of Books*, 28 January 1971.

Thompson, E. P., 'The Moral Economy of the English Crowd in the Eighteenth Century', *Past and Present*, no. 50 (1971) 76–136.

Ward, J. T., 'West Riding Landowners and the Corn Laws', *English Historical Review*, lxxxi (1966) 256–72.

Webb, Sydney, and Webb, Beatrice, 'The Assize of Bread',
  Economic Journal, xiv (1904) 196–218.

Unpublished Material

Lemisch, L. Jesse, 'Jack Tar Versus John Bull, The Role of New
  York's Seamen in Precipitating the Revolution', unpublished
  Ph.D. dissertation, Yale University, 1962.

Index